ELECTRICAL ESTIMATING Work For A Profit

Paul Rosenberg

Prentice Hall, Englewood Cliffs, New Jersey 07632

Library of Congress Cataloging-in-Publication Data

ROSENBERG, PAUL.
Electrical estimating : work for a profit / Paul Rosenberg.
p. cm.

Includes index.
ISBN 0-13-247586-3
1. Electric contracting. 2. Electric apparatus and appliances--Installation—Estimates. I. Title.
HD9716.E432R67 1989
621.319'2'0681—dc19 88-25314

Cover design: BEN SANTORA
Manufacturing buyer: MARY ANN GLORIANDE

A Division of Simon & Schuster
Englewood Cliffs, New Jersey 07632

The publisher offers discounts on this book when ordered in bulk quantities. For more information, write:
Special Sales/College Marketing
Prentice Hall
College Technical and Reference Division
Englewood Cliffs, NJ 07632

Printed in the United States of America

10 9 8 7 6 5 4 3 2 1

ISBN 0-13-247586-3

Prentice-Hall International (UK) Limited, *London*
Prentice-Hall of Australia Pty. Limited, *Sydney*
Prentice-Hall Canada Inc., *Toronto*
Prentice-Hall Hispanoamericana, S.A., *Mexico*
Prentice-Hall of India Private Limited, *New Delhi*
Prentice-Hall of Japan, Inc., *Tokyo*
Simon & Schuster Asia Pte. Ltd., *Singapore*
Editora Prentice-Hall do Brasil, Ltda., *Rio de Janeiro*

CONTENTS

PREFACE

This book is written to explain the art of making profits in the electrical-construction business. The emphasis of the book is placed on the financial aspects of the trade rather than on the engineering and technical aspects. Certainly the technical aspects of the electrical trade are important, but they represent only the first step in the pursuit of profits. Given that a sincere contractor (or future contractor) has the required technical knowledge (and there are a great number of good books on the subject), this book takes the reader through the next step of training as an electrical contractor, going beyond engineering and installation techniques into the fine art of producing profits.

The background of this book comes from close experience with a great number of electrical-contracting firms on both sides of the fence. The contractors from whom this information has been gleaned have been acknowledged to be some of the finest in the business. Some of the individual company owners are among the shrewdest businessmen in the country. These are men who learned how to find the advantages that made their company's work profitable. The methods shown in this book are the factors that will make the difference between a company that just stays busy and a company that makes a profit; one doesn't necessarily follow the other.

One aspect that this book deals extensively with is the understanding of not just *which* techniques can be used to make electrical projects more profitable, but *why* these techniques make the work more profitable. The intent is to explain the hidden laws of profitability so that the reader can apply them to any situation that he or she encounters. Individual techniques are good to know, but as with almost anything, the electrical business is continually changing, and what works at one time may not at another.

Because the electrical market is changing, the book takes a look at the directions in which the market will be heading in the future. How trends develop, how to spot them before they get big, and how to capitalize on them are a few of the topics that are covered.

It is important to understand that the face of electrical construction is changing. In the past, most contractors made their living by simply bidding on new construction projects. They played a type of percentage game, hoping to get every fourth or fifth project they bid on. While this way of doing business worked acceptably for a while, it has changed now in almost all areas. The easing of licensing requirements over the past decade has resulted in a much larger number of registered electrical contractors. Between 1972 and 1982 their number increased by 22 percent, the largest increase of any major trade. Judging by the number of subscribers to trade magazines, the increase continues. Granted, the dollar amount of electrical construction work has also increased dramatically, but a great deal of this increase is attributable to inflation and productivity increases (the ability to do the same job faster due to the use of better tools, materials, and methods). However you look at the statistics, the fact remains that in the town where there were five electrical contractors doing business in 1972, there were six in 1982, and perhaps seven now. The population of the United States has not risen by a corresponding amount, so the number of electrical contractors per capita has risen quickly. It can further be added that there is now a much larger number of inexperienced contractors (inexperienced in a business sense) than in the past. By nature, these new contractors are going to make mistakes that will to some degree hurt the market. Because they compromise a larger part of the market, these influences on the market will be felt more strongly.

For these reasons (and others), electrical contractors who want to maintain or increase their profits from here on will have to take a harder look at the business aspects of their trade. There is, and will be, a need for electrical contractors to intelligently market their services. The intent of this book is to communicate this knowledge to the reader, so that he or she can guide a company into the future without learning lessons the hard way.

Chapters One through Five give detailed explanations of the basic laws and practices of profitability that are shared by all electrical contractors. The problems and practices of selling, marketing, management, planning, and accounting are explained from a profit standpoint. Chapter Five, Cutting Costs, is especially important. It covers all of the major techniques for cutting costs for an electrical-construction company, and also explains how far you can go in cutting costs and why you need to avoid going too far. A large section of the chapter is devoted to purchasing techniques, which is of great significance because materials account for half to two-thirds of the average electrical project.

Chapters Six through Ten take the reader through several different types of electrical contracting, explaining the characteristics of each type. Each type is explained as to its advantages, disadvantages, which types of markets (big city, small town) it works best in, and why. The techniques and "inside tricks" of successful contractors are explained for each classification. These chapters are a how-to guide for service work, residential wiring, specialty work, maintenance, and design/build work. Chapter Eleven is a guide to the professional and general business phases of the trade, such as legal considerations, liability, insurance, investments, and accounting. The last chapter, How The Game Is Really Played, is a look at the real inside dealings that go on in the electrical construction industry, and how to protect yourself and your company.

At the time of this writing, this work is the only published guide to profitability for electrical contractors. While this book does not (and cannot) eliminate the importance of the National Electrical Code, its information is equally important as far as the maximization of profit is concerned. The two are to be used together. Knowledge of the Code and a study of technical books is essential to any electrical contractor as a preliminary step to making money in the electrical industry. If the Code and installation techniques make up the first level of education for an electrical contractor, this book provides a second level which will teach the student how to reach goals.

The electrical industry is laden with both tremendous opportunities and substantial risks. This book provides the information you need to locate the opportunities, and wade safely through the risks.

Best Wishes,

PAUL ROSENBERG

1

WHAT MAKES ELECTRICAL WORK PROFITABLE?

Just what is it that makes electrical work profitable? There isn't any one answer, but in its most elementary form, the answer is the law of supply and demand. Obviously this is quite simplistic, but it is the core of the issue.

SUPPLY AND DEMAND

Basically the law of supply and demand says this: When the demand for a product or service exceeds the available supply, the price of that product or service goes up. If the supply of a product or service exceeds the demand, the price of that product or service goes down. This is a basic principle of economics.

Look at how this applies to the electrical-construction marketplace. Let's go through a typical scenario: In a certain medium-sized city there are twelve electrical contractors who have been doing business for a number of years. Let's assume that the amount of electrical work done in that city is fairly constant from year to year. Now, in one year three separate electricians get their own master-electrician's licenses, and quit their jobs to start their own companies. We can assume that these three are competent electricians, but somewhat limited as far as being experienced at several different types of wiring installations.

So what happens? These independent electricians start calling on general contractors to bid on residential and small commercial projects that they might be working on. Now the situation is that rather than there being twelve electrical contractors competing for the same job, there are fifteen. The supply of the service (electrical contracting) has gone up by 25 percent with no increase in the demand for the service. The price level for residential and small commercial wiring in this town will drop. How much it will drop depends on a number of other factors, which we will study later in the book. The point we want to make here is that supply and demand are tied to each other. What would happen in the same town if one year three of these twelve contractors decided that they wanted to retire? Then the customers would have to go looking for a new electrical contractor to work for them. You would have general contractors calling up electricians asking if they would please take the time to look at a new job they are figuring. The general contractor really wants to get the job, but his old electrical contractor is gone, and he is looking desperately for someone who can do the work for him. To the general contractor in this situation, price loses importance, and finding someone who is able to do the job becomes the important factor. Obviously, the price level in this town will go up, as the supply has dropped by 25 percent with the demand remaining the same.

Regardless of the complexity of the marketplace, this principle still rules. It is up to each electrical-contracting company to position itself in the second of these two situations, rather than the first. Ultimately, this is what is going to make the company profitable, all other things being equal.

The Marketplace Determines Profitability

One of the most common mistakes that electrical contractors make is that they try to sell what they like to do best, rather than what is most profitable. The marketplace is what determines whether one type of work is more profitable than another, and only the marketplace can make this determination. The electrical contractor who goes about selling the service that is easiest to do is often in for a sad surprise. What is easiest and most convenient is usually what goes for the least cost.

The point is that to do profitable work, you must make intelligent, informed decisions as to what types of jobs you are going to pursue, rather than bid on every job that comes before you. To get profitable jobs, the electrical contractor must first understand the principles of marketing. Just getting enough work to be busy isn't good enough; there are plenty of electrical contractors who do this and go broke every year. To make a reasonable profit, you msut get the types of jobs that make money.

In its simplest form, an electrical-contracting company (or any contracting company) can be said to have two basic job functions:

1. Getting work.
2. Doing work.

In our discussions of marketing, we are going to forget about the *do work* function, and concentrate on *getting work.*

Because of electrical contracting's strong technical orientation, the managers and owners of electrical-contracting companies have spent the majority of their time and energies on the *do work* part of their businesses, rather than on the *get work* part. Certainly, doing work well is essential to any electrical-contracting firm, but even though you may be tremendous at doing your jobs well, your profits will be limited by what type of work you get.

MARKETING

Marketing is one of those words that sounds like it would be good for General Motors or IBM to think about, but never really applies to electrical contracting. Well, that simply isn't so. A good marketing plan is important to any company that is interested in maximizing its profits (which is all of us). There is no mystery to the science of marketing. As we said, an electrical-contracting company has two basic functions, getting work and doing work. Marketing is simply the science of getting work in the best possible way, nothing more.

Notice that we are making a big distinction between simply getting work and getting the best type of work. This is a particularly important point for electrical contractors. Typically, an electrical contractor says "I've done my job of getting work; we've got four projects going right now, and can hardly keep up." The man is correct in his statement that he has gotten work, but the important questions are: *Is it profitable work? Is it the most profitable type of work that is available in the area? Is there likely to be any repeat business from this customer?* Many electrical contractors never even consider these questions, and yet they appear to have done their job of getting work.

This is where marketing comes in. Marketing gives direction to your sales efforts; it identifies the most profitable segments of the electrical market and charts a course for obtaining more work of those types. Without some type of marketing plan, a company's sales success is certainly going to be hit-and-miss; and probably miss as much as it hits.

Almost every industry besides the counstruction industry has learned how to best market its products. In fact, the large corporations think nothing of spending millions of dollars for sales and marketing studies. Why? Because it makes a big difference in the bottom line. If you can identify which segments of the market will allow a higher percentage of profit than the others, you can position yourself to get more of that type of work. If you know that a certain part of the market will not allow for a good percentage of profit, you can stop trying to get that type of job. The whole concept is very simple, but the fact is that very few electrical contractors spend any appreciable amount of time in making a sales and marketing plan.

Developing a Marketing Plan

Why develop a marketing plan? For the same reason a driver in a cross-country race takes a road map. If he wants to win, he has to know the best routes. Without a roadmap, he might get there, but he is certain to take plenty of wrong turns along the way.

In Table 1–1 we show a list of the most important marketing questions to ask yourself as you develop a plan for your sales efforts. A marketing plan should be made at least once every year. Your plan should spell out exactly what volume of sales you want in the next year and what type of sales you want. Table 1–2 shows a typical sales projection, which

TABLE 1–1 Questions for Sales and Marketing

1. What does our market look like for next year?
2. What are the long-term forecasts for the markets we are in?
3. Who are our chief competitors?
4. What are their strengths and weaknesses?
5. What are our strengths and weaknesses?
6. Should we look into a different type of market? (Contract-bid work, design/build, service work, specialty work, etc.)
7. What will it take to get into these markets?
8. Are there any profitable new markets opening up?
9. What is our most profitable type of work?
10. Can we get more of these types of jobs?
11. What are our least profitable types of jobs?
12. Do we really need them?
13. Who are our best customers?
14. Are they happy with the services they receive from us?
15. What are their most common complaints?
16. How can we correct them?
17. Do we need to find some new customers?
18. What is it that customers really want from an electrical contractor?
19. Can we provide them with a service that is better than what our competition provides?
20. How can we best reach customers?
21. What will influence customers to choose us over another electrical contractor?

TABLE 1–2 Sales Projection

	Sales	Sales Applied Toward Overhead Profit	
Contract-bid work (Bonding limit)	$1,000,000	18%	$180,000
Design/build work (With Roberts & A. James Construction)	$250,000	24%	$60,000
Service Work (Repeat Customers and ad response)	$150,000	33%	$50,000
Residential Work	$100,000	33%	$33,000
Contract Maintenance Work (Argus Axle)	$20,000	30%	$6,000
Total	$1,520,000		$329,000

is the result of a planning session. You will notice that each category of work has its own total sales figure and its own overhead-and-profit percentage figure.

In such a planning session, the company's key personnel should discuss the questions that were shown in Table 1–1 and work together toward making a solid plan of attack for the following year. Make sure also that you prepare well for your planning session. Do a little research, and analyze the state of your local market. This way, you have a starting point for your meeting—a good data base and recommendations which can be discussed, modified, or adopted. Even if your final plan looks nothing like the original recommendations, you will have a data base to work from, which is very important.

Analyzing the Marketplace

For electrical contractors, making a good market analysis is not a cut-and-dried proposition. Sales figures for electrical contracting are virtually non-existent; those that are available are usually a year or two behind and only cover the national and regional markets, which makes their value to most electrical contractors only marginal. So to make a reasonably accurate market analysis, it is necessary to gather all the pertinent information that can be found, carefully review each piece of information, and slowly consolidate all of the facts to get a good picture of the market.

The first place to start is your own company. Go through your records for the previous year to see what types of jobs came in over budget most often, what types of jobs were the most profitable, which foreman had the most profitable jobs, and so on. If you keep an estimate log (a written record of estimates that your company did—both jobs that you got, and jobs that you didn't get—you can figure out your success rates for various types of jobs. For example, if you find out that you get one out of every four commercial jobs that you figure, but only one out of every seven industrial jobs that you figure, you might decide to concentrate more on commercial work than on industrial work for the coming year. Another interesting thing you can do with an estimate log is to figure the costs of doing your estimates. If you determine how many hours the estimators have spent on various types of jobs, you can figure how much it is costing your company to prepare estimates for the different types of jobs. On the average, these costs come out to between $1300 and $1700 for each million dollars of estimates which are produced. Industrial jobs are a little higher than other types, and residential and small commercial jobs are typically a little lower. The larger the job, the lower the cost per million; the smaller the job, the greater the cost per million.

One other thing to remember when analyzing estimating costs: In most electrical companies the estimator does a lot more than just prepare estimates; he or she typically plays the role of superintendent and project manager. It wouldn't be fair to expect a full service estimator to crank out a million dollars worth of estimates for every $1500 that the company pays him, since he does a lot more than just estimate.

As you go through your company's records and analyze the figures, make sure that you write down everything of value. You will need it later; it helps a lot to have everything down in black and white in front of you.

The second thing to do is an analysis of what the other electrical contractors in your area are doing. What types of work does each one specialize in? Are they doing well or having trouble? What do you expect each of them to do next year? As with all of your market-analysis information, write everything down. Take into account any factors that would affect your competitors.

Next, try to get a feel for what the market will be like in the next year. One of the best ways to do this is to call several architects and engineers in your area to see what types of projects they are working on. Don't ask them specific questions like the names of the owners, but try to find out what types of jobs they are working on (commercial, industrial, schools, hospitals) and how big the jobs are. Be sure to use any good contacts that you have. You can often get some good information from a local Chamber of Commerce.

Trends in the Marketplace

Keeping up with new trends in the marketplace is one of the most profitable things that an electrical contractor can do. It is important for even a small contractor to stay in touch with new trends, because in just a few years these trends will work their way from the largest cities down to the small towns and the countryside.

We are all familiar with the great increase in communications that has occurred in the last five or ten years. What this means to the electrical business is that new trends are going to work their way through the industry much faster than they ever have before, which makes it that much more important to stay in touch with the new trends.

The things to look for are new types of electrical systems, new types of installation techniques, new types of tools, even new management methods. The computer-power market is a classic example of how trends develop. If you can remember back a number of years, you will recall a lot of discussion of the subject in trade magazines. Even before there were technical articles about computer power, there were clear signs that in a few years computer power would be a big market for electrical contractors. Many smaller contractors more or less ignored the subject, and in so doing passed up a lot of good, profitable work. Even in some of the largest cities, it was very hard to find an electrical contractor who really understood the problems of getting a good high-frequency ground. As late as the mid-1980s computer salespeople in some of the largest cities were frustrated trying to find electrical contractors who could understand what they wanted to do (although in all fairness, some of the computer salesmen weren't too sure what they wanted to do themselves). A knowledgeable electrical contractor could have gotten some very profitable business, as he would have been not only a contractor, but a trusted consultant also. This was very typical of the early 1980s, and is still true in some of the smaller markets.

The really sad part of the story is this: A knowledgeable electrical contractor in this situation has a tremendous demand for his services and virtually no competition. Remember what we said about supply and demand: When the demand is up and the supply is low, the price goes up. A contractor who could show that he was knowledgeable and experienced in computer power could practically name his price. Those contractors who took some time and effort to familiarize themselves with it were paid handsomely for their foresight.

The thing to remember about this is that the signs were there but not very many people took advantage of them. The real core of the problem is that most people in the electrical-contracting business don't like to do things unless they are unavoidable. It is easy for them to leave new trends alone until they are forced to deal with them. This is why it is so important to always look to the future. The vast majority of electrical contractors are so wrapped up in taking care of their business's day-to-day operations that they never take any time out to plan for some long-term improvements. The signs of the future are almost always there to be seen, so if you would like to be in a better position in a few years than you are now, all you need to do is to keep your eyes open and take some initiative, regardless of whether it is convenient.

There are many different types of trends developing all the time, with the various types of trends mattering more or less to different types of contractors.

Regional trends. Regional trends are of interest primarily to the largest electrical contractors, but should be watched by smaller contractors also. One recent regional trend has been the population shift from the north and east to the south and west. Obviously, there will need to be a marked increase in the number of residential housing units that will be built in these areas, which will be followed in a few years by new schools, hospitals, prisons, road construction, and utility construction.

A large contractor may do some research and then decide to open up a branch office in one of these areas, usually in the largest city in the region. Then the contractor will specialize in one or two types of projects that he thinks will be plentiful in the particular area.

Maybe the contractor has a number of people who are highly experienced at wiring hospitals. They would set up shop in a high-growth area, bid every hospital job that comes along, and very little else. Because of their location in this type of area, they can get enough hospital jobs so that they don't have to do much else. Because they specialize in one specific type of work, they can do it much more efficiently than a contractor who doesn't specialize in that type of work. Because of this, they can bid lower than most other electrical contractors on this type of job and yet make a better profit. This technique has worked very well for the very large electrical contractors.

In the same way, but on a smaller scale, this is important to a smaller contractor also. If they can ascertain what types of work will be increasing or decreasing in their particular area, they can fit themselves to the expanding segments of the market in general, and thereby get more and more profitable work. The key is to stay at least one step ahead of the new trend. If you can be ready when the trend actually starts to hit, you will almost always be several steps ahead of your competition. You will be in the position to be one of the few suppliers when the demand is at its highest. On the other hand, if you wait until the trend hits and then react, you will just be another face in the crowd. Instead of being in demand, you will remain a generic electrical contractor.

Market trends. Market trends reflect a change in the type of work that is done. The computer-power market is an example of this type of trend. With the growth of computers has come a new phase in the electrical-contracting industry. It has literally affected every marketplace in the United States, and will probably change the electrical-contracting industry throughout the entire world by the turn of the century. But the trend is far from over; rather it is evolving into several different trends. Commercial buildings, factories, and even residences are being wired with complex, integral electronic systems at an increasing rate. UPS systems (Uninterruptable Power Sources) are becoming far more common, and cogeneration is increasing in importance. Electrical contractors are being called upon to perform these installations, and they should be spending considerable effort learning how to do these installations well.

The breakup of AT&T was a very important change that electrical contractors should have paid attention to, but regrettably, not very many did. The telecommunications market has been growing by leaps and bounds, but most electrical contractors have let it go to small specialty contractors rather than expending the time and effort to learn how to do this work and set up shop.

It seems likely that by the turn of the century the electrical-contracting business will do almost as much electronic contracting as it does electrical contracting. However, unless a lot of electrical contractors have a change in attitude there will be a large number of specialty-electronics contractors, and a smaller number of old-fashioned pipe-and-wire electrical contractors.

Trends in installation equipment. Now we come to trends in different methods of doing installations. Usually this means new types of tools and techniques. Every year new tools and equipment are introduced to the industry. The only sales pitch that the manufacturers can make for these products is that they make electrical installations either easier or faster, so they are worth noticing. Some of the more important new items in recent years have been cordless drills and screwdrivers, wire carts and wire sold in spools rather than in boxes, motorized scaffolds, and so on. These products have a noticeable effect on productivity. Again, the news is there to be seen if you take the time to notice it. Almost every significant new product receives attention in the trade magazines.

Management trends. Finally, there are trends in the management of electrical-contracting firms. Twenty years ago it was rare to find electrical contractors who had their own computers; but now, about three-quarters of the large contractors and one-half of the medium-sized contractors have their own computer systems. A lot more contractors are spending some time learning about business. Many take evening classes at community colleges. The percentage of electrical contractors who have been to college is much higher now than it was even ten years ago.

There are definite changes taking place in the way electrical contractors manage their businesses. Your competitors are changing, and it is important for you to know how they are changing.

Analyzing Your Local Market

There are a number of techniques for analyzing a local marketplace, or even a certain segment of it. Table 1–3 shows a forecast for electrical contracting sales in 1987. These figures have been taken from a number of sources and consolidated into one total sales figure for the 50 United States, the District of Columbia, and 121 separate metropolitan areas. If you live in one of these metropolitan areas (and most people in the United States do) you can get a good starting point to figure out just how much work is available in your particular area.

It is important to note that these figures are for electrical contractor's total sales, and include materials, labor, overhead, and profit.

Market mix percentages for all fifty states and the District of Columbia are shown in Table 1–4. These percentages are very helpful for figuring out exactly how much work of each type is available for each particular location. Remember though that these figures are averages, and that your particular marketplace may be somewhat different from the average. Each marketplace has its own characteristics and peculiarities. Even different marketplaces within a certain region can be vastly different. The percentages are shown for residential work (all types, including multi-

TABLE 1–3 Electrical Contractor Sales Forecast—1988

Market Area	Expected Sales (in millions)
Alabama (whole state)	$528,680
Birmingham	189,809
Mobile	94,318
Alaska	80,792
Arizona	366,744
Phoenix	254,031
Tucson	85,922
Arkansas	207,932
Little Rock, North Little Rock	125,548
California	4,024,020
Anaheim, Santa Ana, Garden Grove	513,099
Bakersfield	50,823
Fresno	97,977
Los Angeles, Long Beach	1,671,516
Oxnard, Simi Valley, Ventura	72,521
Riverside, San Bernadino, Ontario	216,973
Sacramento	92,582
San Diego	195,895
San Francisco, Oakland	719,390
San Jose	276,640
Stockton	47,152
Colorado	489,510
Denver, Boulder	390,139
Connecticut	716,431
Bridgeport	114,910
Hartford	250,120
New Haven, West Haven	74,540
Stamford	50,119
Delaware	83,432
Wilmington, Delaware, New Jersey, Virginia	49,434
District of Columbia	248,609
Washington, D.C., Maryland, Virginia	708,790
Florida	2,299,170
Ft. Lauderdale, Hollywood	257,400
Jacksonville	154,921
Lakeland, Winter Haven	47,147
Miami	1,049,400
Orlando	244,100
Tampa, St. Petersburg	591,240
West Palm Beach, Boca Raton	169,610
Georgia	1,210,500
Atlanta	809,842
Hawaii	95,400
Honolulu	67,100
Idaho	105,252
Illinois	1,150,792
Chicago	912,942
Peoria	35,112
Rockford	34,924
Indiana	554,441
Evansville	45,858
Ft. Wayne	39,060
Gary, Hammond, East Chicago	40,960
Indianapolis	181,421

TABLE 1–3 Continued

Market Area	Expected Sales (in millions)		Expected Sales (in millions)
Iowa	263,520	Missouri	662,544
Davenport, Rock Island, Moline	42,826	Kansas City, Missouri, Kansas	585,634
Des Moines	70,848	St. Louis	394,784
Kansas	564,005	Springfield	29,421
Wichita	85,248	Montana	53,174
Kentucky	375,161	Nebraska	138,240
Lexington, Fayette	36,558	Omaha	75,024
Louisville, Kentucky, Indiana	121,064	Nevada	127,751
Louisiana	580,117	New Hampshire	112,598
Baton Rouge	62,021	New Jersey	1,130,335
Lafayette	39,684	Jersey City	84,705
New Orleans	224,784	Long Branch, Asbury Park	62,625
Shreveport	84,654	New Brunswick, Perth Amboy, Sayreville	320,115
Maine	178,173	Newark	420,285
Maryland	632,601	Paterson, Clifton, Passaic	92,100
Baltimore	338,690	New Mexico	146,089
Massachusetts	1,130,940	Albuquerque	57,289
Boston	816,655	New York	2,922,150
Springfield, Chicopee, Holyoke	80,434	Albany, Schenectady, Troy	91,815
Worcester	86,615	Buffalo	176,910
Michigan	985,766	Nassau, Suffolk	764,010
Detroit	554,901	New York, New York, New Jersey	1,498,560
Grand Rapids	90,570	Rochester	162,870
Lansing, East Lansing	40,650	Syracuse	131,040
Minnesota	514,282	North Carolina	1,090,501
Minneapolis, St. Paul	320,817	Charlotte, Gastonia	294,729
Mississippi	232,724	Greensboro, Winston-Salem, High Point	196,451
Jackson	58,692	Raleigh-Durham	144,067
		North Dakota	122,400

Ohio	1,327,250
Akron	60,554
Canton	47,500
Cincinnati, Ohio, Kentucky, Indiana	207,100
Cleveland	425,721
Columbus	220,179
Dayton	91,838
Toledo	86,804
Youngstown, Warren	56,954
Oklahoma	201,574
Oklahoma City	77,071
Tulsa	69,583
Oregon	515,970
Portland	331,451
Pennsylvania	1,379,655
Allentown, Bethlehem, Easton, Pennsylvania, New Jersey	74,685
Harrisburg	34,185
Lancaster	39,780
Philadelphia, Pennsylvania, New Jersey	589,045
Pittsburgh	244,770
York	41,385
Rhode Island	120,638
Providence, Warwick, Pawtucket, Rhode Island, Massachusetts	125,597
South Carolina	459,461
Charleston, North Charleston	85,471
Columbia	91,355
Greenville, Spartanburg	145,470
South Dakota	70,848
Tennessee	940,240
Chattanooga	110,376
Johnson City, Kingsport, Bristol	69,471
Knoxville	110,332
Memphis, Tennessee, Mississippi, Arkansas	353,174
Nashville, Davidson	196,662
Texas	2,353,100
Austin	88,921
Beaumont, Port Arthur, Orange	48,574,
Corpus Christi	59,302
Dallas, Ft. Worth	765,860
El Paso	74,947
Houston	700,300
Lubbock	36,771
Odessa	38,441
San Antonio	142,146
Utah	162,010
Salt Lake City, Ogden	104,370
Vermont	75,968
Virginia	913,879
Norfolk, Virginia Beach, Portsmouth	172,490
Richmond	207,835
Washington	562,183
Seattle, Everett	429,101
Spokane	58,768
Tacoma	68,541
West Virginia	187,246
Charleston	52,027
Wisconsin	309,320
Madison	31,859
Milwaukee	141,770
Wyoming	77,689

TABLE 1–4 Electrical Contracting Market Mix—1988

AREA	% Residential	% Commercial	%Industrial and Institutional
Alabama	34	32	34
Alaska	33	31	36
Arizona	41	34	25
Arkansas	34	33	33
California	36	36	28
Colorado	40	36	24
Connecticut	27	33	40
Delaware	31	33	36
District of Columbia	32	32	36
Florida	34	32	34
Georgia	32	32	36
Hawaii	35	36	29
Idaho	41	35	24
Illinois	16	33	51
Indiana	17	34	49
Iowa	38	33	29
Kansas	37	34	29
Kentucky	32	31	37
Louisiana	24	36	40
Maine	27	33	40
Maryland	32	32	36
Massachusetts	27	34	39
Michigan	16	32	52
Minnesota	38	34	28
Mississippi	34	32	34
Missouri	38	33	29
Montana	40	37	23
Nebraska	38	33	29
Nevada	42	35	23
New Hampshire	27	33	40
New Jersey	23	40	37
New Mexico	42	35	23
New York	23	40	37
North Carolina	32	32	36
North Dakota	39	34	27
Ohio	17	32	51
Oklahoma	23	36	41
Oregon	37	35	28
Pennsylvania	25	38	37
Rhode Island	27	33	40
South Carolina	32	32	36
South Dakota	37	33	30
Tennessee	33	32	35
Texas	24	36	40
Utah	41	35	24
Vermont	27	34	39
Washington	37	35	28
West Virginia	30	32	38
Wisconsin	16	33	51
Wyoming	40	36	24

family), commercial work, and industrial and institutional work. Tables 1–3 and 1–4 work together. For example, a contractor who works in Cleveland, Ohio could refer to Table 1–3 and find out that the total electrical-contractor sales in his area are expected to be $425,721 million; then he could go to Table 1–4 and find out that about 17 percent of that work will be residential, 32 percent of the work will be commercial, and the remaining 51 percent of the work in his area will be industrial and institutional. So the total volume of residential work in his area will be approximately $72,373 million, the total volume of commercial work in his area will be about $136,231 million, and the total volume of industrial and institutional work in the same area will be about $217,117 million.

Another method of calculating sales potential is shown in Table 1–5. Using this method, it is possible to estimate sales for very specific segments of the market.

The first category shown is Institutional; this includes schools of all types, hospitals, prisons, and the like. The figure shown for total electrical sales is $180 per employee per year; this includes material, labor, and markup for overhead and profit. You will notice, though, in note 1, that only about half of this work is done by outside electrical contractors, with the other half done by in-house maintenance people. Because of this, a

TABLE 1–5 Electrical Sales Figures

Market Segment	Average Yearly Sales
Institutional	$180 per employee[1]
Industrial	$550 per employee[2]
Commercial	$ 50 per employee[3]
Residential	$ 30 per household[4]

[1]Only about one-half of institutional electrical work is thought to be done by outside contractors. Therefore a better figure for electrical contractor's sales would be $90 per employee, per year.

[2]Only about one-third of industrial electrical work is done by outside contractors, so a better figure for electrical contractor's sales would be $183 per employee, per year.

[3]Approximately three-quarters of commercial electrical work is thought to be done by electrical contractors, so a better figure for electrical contractor's sales would be $37.50 per employee per year.

[4]The percentages of residential electrical work done by electrical contractors vary greatly from one community to another. Generally, between 50–80% of residential electrical work is thought to be done by outside contractors.

more accurate figure for electrical contractor sales is $90 per employee per year. Notes 2, 3, and 4 to this chart make similar adjustments to the industrial, commercial, and residential figures that are shown.

For example, if you want to figure out how much institutional work is available in your area, multiply $90 times the number of employees of all the institutions in your area. For industrial work, multiply $183 times the number of people in your area that are employed in industry. For commercial, multiply $37.50 times the number of retail employees in your area. To figure how much residential work is available in your area, multiply the number of households in your area by about $20; you will have to adjust this figure up or down a little bit depending upon your area.

Before you can do this, however, you need to have the figures for employment and number of households. While these figures are somewhat unusual, they are not too hard to obtain. The easiest way to get these figures for any particular area is by simply calling the research librarian at a local library. It is the researcher's job to find answers to these types of questions, and they usually do a pretty good job of it. The employment figures are put out by the U.S. Bureau of the Census in a publication called *County Business Patterns,* which is published annually. There is one thing to remember about these figures though—they are two to three years behind, due to the time it takes to compile that much data. If you would like to get information about a specific firm, you can find these figures (for large companies only) in a publication called *Standard Corporate Descriptions,* which is published by Standard and Poors, or in Dun and Bradstreet's *Million Dollar Directory.* The number of households for a particular area is published annually by *Sales and Marketing Management* magazine in their "Survey of Buying Power" issue.

The last step in analyzing your local marketplace (which is essential) is simply to talk to people in the industry. Information from people in your local area is the only really specific information that you can get, as most of the other information that is available to you will be based upon national averages rather than your local area.

The best people for you to talk to are electrical distributors and other electrical contractors. Electrical distributors can tell you a lot about what is going on in the area, but don't try to get them to give you specific information about individual contractors, as it would not be ethical for them to do so. One of the easiest ways for electrical distributors to lose customers is for them to betray their customer's trust, so don't ask them to do it. Architects and engineers are also good for giving information about upcoming projects, but the information that is available from them is only for large projects, and cannot account for residential work, maintenance, service work, and the like. The electrical distributors have a first-hand view of the entire electrical marketplace; they know what types of work are being done and how they are being done. No one else can tell you as much about the local marketplace as they can. Most distributors will be quite happy to talk to you, as long as you don't try to get them to talk about specific customers.

Electrical contractors are another story, as most electrical contractors don't trust other electrical contractors enough to give them information. Because of the competitive marketplace and all of the pressures that stem from it, many electrical contractors fear that any information they give may be used against them (and rightfully so). If you have other electrical contractors that you can cooperate with, you can help each other quite a bit. Typically though, this only works for contractors who do different types of work. Sometimes a contractor's association can work well this way. Cooperation between electrical contractors can be beneficial for all involved, but it can be pretty difficult to foster sometimes, and once someone betrays another's trust, it is very hard to restore.

To really make a good analysis of your market, you will have to integrate your figures with these interviews. Both are important, but you need them together to get a good picture of what is really going on in your market.

Choosing Market Segments

After you have gotten all of the pertinent information together, you will have to review it and make some decisions as to which segments of the market you want to work in. The critical factors in making decisions are these:

1. What types of work are the most profitable for your company.
2. What types of work look like they will be the most plentiful in the next year.
3. What types of work look like they will be the most plentiful in the years to come.
4. What types of work your competitors do best.
5. What you expect your competitors to do in the future.
6. How easily your company can adapt to doing different types of work.

Choosing market segments is a very important decision for any electrical contracting company, so go through a slow, careful analysis before making any firm decisions. In your planning, be sure to consider not only the next year, but make some long-range plans for two years, five years, and ten years. Of course these long-range plans will change over the course of time, but it is better to have a plan to work from, rather than no plan at all.

Being in the Right Place at the Right Time

Being at the right place at the right time is the essence of what it takes to make the good profits in the electrical business. Certainly there are a lot of other "musts" (being organized, doing your work well, working hard, etc.), but without being at the right place at the right time, the rewards will still be meager, even though you do your "musts" well.

This is why marketing is important. Marketing is the business science of being at the right place at the right time. The purpose of a marketing survey is to indentify where the right place to be is, and when to be there. It makes a difference—sometimes a big difference. Profit margins in the high supply–low demand and the high demand–low supply parts of the market can vary by 20 percent or more. At a volume of $400 thousand per year, that makes a difference of $80 thousand in the owner's pocket, which makes being in the right place at the right time very important indeed.

Every now and then you hear someone say, "It was just luck that so and so made it big; nothing more than pure chance that he happened to be in the right place when the market broke loose." While there is no question that things often happen that are pure chance, you will notice that these types of statements come most commonly from someone who was not in the right place at the right time. Certainly there is a factor of chance, but most of the time there is a lot more than chance that comes into play. The fellow who is in the right place when the market breaks is seldom there by accident. He is usually the one who worked a little smarter and harder than the others. As one old-time contractor put it, "The harder I work, the luckier I get."

2

MANAGEMENT AND PLANNING

Managing your company well is critical to its success. It is management's job to keep the company working together; without this, no company can survive. The days are gone when a contractor could manage his company in a haphazard style and survive; in the modern market managing your company is not optional, it is essential.

THE BASICS OF MANAGEMENT

To illustrate the basic principles of management, we start by comparing your company to a human body. In order for your body to do anything, it must first get directions from your brain. In the same manner, a company should operate on directions from its management.

Management's job is to direct the company's operations; to give the company the directions that it needs to operate. Obviously this is a lot easier in principle than in practice, but with some time and effort, it can be effectively accomplished. A key point to remember is that a company's management must not only give correct directions to the rest of the company, but it must give *enough* directions to the rest of the company. Just as your body needs to activate a number of muscles in unison in order to move; your company needs to have each of its parts working together. If the proper signals are not sent at the proper time, the result will be confusion. This is the reason that the management of most electrical contracting firms spends so much of its time fixing problems. If they could set up a coherent system of managing the company's operations, the amount of time that is spent resolving crises would be cut in half.

Management's First Job

The first job of management is overall planning, which should be done at least once per year, preferably at the beginning of the fiscal year. The purpose of overall planning is to carefully review the fundamentals of a business's operations and profitability. In Table 2–1 is shown a list of topics that should be covered in an overall-planning meeting. In these sessions you should go over each of the topics listed, taking care to emotionally detach yourself from the company as much as possible so that all of your decisions are based upon facts rather than upon hopes and emotions.

For such meetings it will be essential that you have your most recent profit-and-loss statements, final reports on all significant jobs completed in the last year, and any other significant papers. Be sure to take good notes, and make a permanent record of the meeting. This way you

TABLE 2–1 Overall Planning Meeting

What are our assets?
What are our liabilities?
What does the market look like for next year?
What are the long-term forecasts for the markets we are in?
Should we start repositioning ourselves into different markets?
Who are our chief competitors?
What are our competitor's strengths and weaknesses?
What are our strengths and weaknesses?
Can we make a profit operating the way we are now?
How much profit did we make last year?
Did our overhead come in within budget?
Was last year's budget correct?
Where were we over or under?
Were our labor rates what we expected?
What will they be next year?
Was our labor productivity as good as we expected?
How can we improve our productivity?
Should we look into a different market? (Contract-bid work, design/build, maintenance, service work, specialty work, etc.)
What will it take to get into these markets?
Are there any profitable new markets opening up?
Who are our key employees?
Are they happy working with us?
What can we do to assure that they will stay with us?
Who are our best customers?
Are they happy with the service they receive from us?
What are their most common complaints?
How can we correct them?
Do we need to find some new customers?
How can we promote our services?
What is our most profitable type of work?
Can we get more of these types of jobs?
What are our least profitable types of jobs?
Do we really need them? What were our chief problems last year?
Why did they happen?
How can we prevent them in the future?
Are our workers' attitudes good?
What can we do to improve their attitudes?
Do our office people have a good attitude?
Are we handling our paper work efficiently?
What improvements can be made?
Are we giving our company enough supervision?
Is anyone overloaded with work?
How can we correct this problem?
Is there anyone who doesn't have enough to do?
How can we correct this?
Will we need more people to implement our plans next year?
Is our management and supervision system working well?
Where can improvements be made?
Are there any personality conflicts in the company?
How can they be corrected?

TABLE 2–1 Continued

Have we had problems with late pay or bad debts?
Should we change our credit and collections policy?
Are there any new laws that we need to consider? (Tax, laws, employee regulations, lien laws, etc.)
Do we need to change our financial structure?
Do we need more or less insurance?
Do we have sufficient credit to operate well?
What are our long-term plans?
Should they be updated?

can refer back to the notes in the future to see how successful you were at accomplishing the things you planned. As you decide during the meetings on things that you will need done, make sure that you also decide who will be chosen to do them; you will save a lot of time this way. If you can conduct these sessions somewhere away from your office, your results will be better, as you won't be burdening your mind with day-to-day problems. Having these meetings out of town would be the best (which is why the large corporations do so), but this is obviously not going to fit into a smaller contractor's budget.

These meetings are the most important thing that you will do for your business all year, so take them seriously. Spend as much time preparing for them as is necessary. The heads of many large corporations often take weeks or even months to prepare for these types of meetings. They chart your company's course for the next year, so they deserve your undivided attention.

Because of the importance of these meetings, the people that attend them should be the best that you can assemble. Certainly one or two people could go through the meetings by themselves and cover all of the material, but the final results would not be nearly as good. The reason is this: Unless your opinions about everything are correct (and whose really are?), there will be no way for you to get a better opinion. With a number of good people present, you can get a number of informed opinions, many of which you may not have thought of yourself. The most trusted people in your company should all be at these meetings, even if they are family members. You wouldn't want your competitors to be at these meetings, but are there any other contractors (perhaps from a different town) that you would trust enough to have in attendance? Other people's opinions can be very important to your company. This is why large corporations have a board of directors rather than one or two people who make all of the decisions. Often, many of the people on the board of directors are not even employees of the company, but are considered to have good business sense, and their opinions are valued. Doing something like this (although somewhat unusual to most electrical contractors) would be to your benefit.

A System of Operations

The next job of management is to set up and maintain a system of operations for the company to follow. This is essential, because without an organized system of what will happen, and how, the result will be widespread chaos. The task of setting up and maintaining a system of operations is not something that can be put off or ignored. Every time something is forgotten or doesn't get done, the company is harmed. The short-term consequences of losing money are bad, but the long-term consequences of having your company's reputation marred are far worse. Losing a good reputation can cost you money for many years to come.

Without a carefully thought-out system, these types of mistakes are sure to be common. The heart of the matter is this: If you want to have a top-notch company, you are going to have to keep it organized and see to it that it runs like a top-notch firm. Unfortunately, most things in life don't just work themselves out; you are going to have to take and maintain control of your organization.

Very simply put, the basic principle is this: You have to identify each thing that must be done, decide who will do it, how it should be done, and then make sure that it actually gets done. The job of making sure that each individual operation gets done may sound impossible, but with a well designed system it can be reasonably well accomplished. The following is a list of the main objectives an electrical-contracting firm's management should address.

1. Control every critical facet of operations, yet remain absolutely as simple as possible.
2. Organize the company's office structure to clearly define each individual's responsibilities. This eliminates confusion in the office.
3. Streamline the flow of paper work, so that all invoices, time cards, purchase orders, and the like go where they need to go as quickly as possible.
4. Provide for all planning, purchasing, and scheduling for each new project.
5. Assure that everything that is supposed to be done actually gets done.
6. Operations should be carefully designed so that they work as well in real life as on paper.

In the last several years we have heard stories about companies that have greatly increased their efficiency by going to a computer system. An interesting fact is that three-quarters of the benefit that is gained from computerizing is not due to the computer system itself, but due to the company organizing itself for the computer system. Several years ago

Westinghouse did a study to determine just how much benefit was obtained by companies computerizing their operations. They found that 75 percent of the benefit had been obtained *before* the computers themselves had ever arrived.

One important point to remember about a management system is that each important task should be delegated to a specific person or department. Human nature is such that it is easy for most people to ignore an unpleasant task, even if it is quite important. So a good system must be designed to be inherently self-checking.

Leadership

The next major job of management is to be leaders. As was said previously, management's role can be compared to that of the human mind controlling the body. As leaders, you must direct the activities and policies of your firm. You are the rule maker. (Not because you want to be better than others, but because it is essential that someone be the rule maker, and as management that is *you*.) You must get the best information that you can. Make a plan of what you want to do, then implement your plan. It is important to remember that this is an essential function of any unified organization. In order for all individuals to work together, there must be a single standard that all work together to achieve. As a leader, you must have a clear vision of exactly where you want the organization to go and how you plan on getting it there. Then you must plant this vision in the others in the company so that all have a single goal to work for. This is the essence of good leadership. Ideally, every individual in the company should have the same idea of where the company is going, and be enthusiastic about getting there.

Cultivating and maintaining this type of attitude in your firm is a never-ending job. You must continually keep in touch with your people's attitudes, plant good attitudes, and stop problems before they spread. Attitudes are volatile and contagious, and as far as productivity is concerned, attitude is almost everything. An employee who just doesn't care will do mediocre work at best, no matter how capable he or she may be. Conversely, if you can get a group of people stirred up about reaching a goal, or overcoming some obstacle, they can do the seemingly impossible.

One problem frequently encountered by leadership is resistance to change. Many people don't like things in their organization to change. They are comfortable in their position as it is, and see a change in the company as a threat to their comfort and stability. Dealing with this can be tricky, and may require some difficult decisions. If the person is very valuable to the company and would be very upset by the change you are planning, you will have to be a diplomat. You could leave their position unchanged and work around them, or you could do something like this:

Call the person into your office, shut the door, tell them what you are going to do, and why you need to do it. Then tell them that you really appreciate the fine job that they have been doing, and that you need their capabilities for this new position or activity. Lastly, ask them not to tell anyone else about the change until you let the news out. Handling changes this way will make employees feel that their positions have been improved, rather than threatened.

As was just illustrated, it is beneficial to consider your employees' feelings, but remember, you can't let one employee stand in the way of the whole company. There must be one clear source of leadership, not many. If you have employees who have a problem with what you are doing, it would be preferable to explain your decision to them and try to keep them happy, but you cannot let them stand in the company's way. Management must dictate policy. If a single employee can force the company management to alter its course, that employee is completely ruining the continuity of the company's leadership. Fix the situation if you can, but don't ruin the company to satisfy one unhappy employee.

Employees

As management, you must implement your company's policies through the people your firm employs. After formulating your policy, you will have to take an honest look at the people who are going to carry it out. Are they capable of doing what you are asking them to do? Maybe you are capable of doing it, but are they? Will they need some type of instruction before they will be able to do the job? Will you need to alter your policy to fit your employees' abilities?

Good employees are few and far-between, as most of us have found out. To complicate this, in the construction industry there is the additional problem of losing the good employees. If a certain employee is doing a good job for the company, it usually doesn't take very long until other people in the industry find out about it. When they need someone who is good, they will often start by seeing if *your* worker is interested.

Always try to think of your firm's employees as important assets to your company, because that is exactly what they are. Without them your company would be crippled. Certainly, a problem employee can make more problems than he eliminates, but fortunately, this type of person becomes apparent quickly.

Try to keep all of your employees (especially key people) happy to be working for your company. Try to keep your workforce stable, especially your office people. If you can do this, you will be far better off in the long run. Over the course of time you are bound to lose some people and take on others, but try to keep turnover to a minimum, and hire the best possible people that you can when you need them.

The company's purpose. One very important thing to remember is that your employees need to know what the company's purpose is and how they will benefit by helping the company to attain that purpose. Your people need to know *Why do we do what we do?* and *What will I get out of it if the company does well?*. These two points are elementary, but they are the things that give your people a reason to get out of bed in the morning, and to show up at the office with a little bit of ambition. Let's go through both of these points.

First, *Why do we do what we do?* Obviously the answer is to make money, but what about some particulars? What are some of the other goals that you would like the company to reach? How about *We want to be the number-one service contractor in the city*, or *We want to do two-million dollars worth of work this year*. As previously stated, your people need a common goal to reach for. There is a vast difference between an employee who is working to help the company reach an explicit goal and an employee who works because the boss will holler at him if he doesn't.

Secondly, *What will I get out of it if the company does well?* This is a decision that the management of each company must make. The answer to this question will vary from company to company and from one employee to another. The important thing is that each individual has something to look forward to receiving when the company reaches its goal (something that the employee considers worthwhile). A profit-sharing plan is one benefit that many companies offer to their employees. Some may offer a bonus or a raise. Exactly what the benefit is doesn't really matter, as long as your employees feel like you will personally benefit by going the extra mile for the company.

Often it works out well to give a key employee a percentage of the company's profits at the end of the year. If you decide to do this, remember one thing: Once you give a certain percentage, it will be expected every year. Cutting this percentage without offending the employee is almost impossible.

Working together. Remember our analogy of management being like the head of the body: The head by itself can go nowhere; it is dependent upon the rest of the body to move it where it wants to go. This is why you should think of your employees as assets, and take a great deal of care cultivating their skills and attitudes. You are dependent upon them to get the job done. Unless you run a one-man shop, you just cannot do everything by yourself; there aren't enough hours in the day.

Remember that ideally the company should operate as one cohesive unit with one common set of goals, working together toward those objectives. One of the best tools that you can use to cultivate this type of attitude is the conference. Sometimes you might want to have a conference for everyone who is involved with a certain job, and sometimes you might want to hold a full company meeting. Someone from management should always run these meetings to keep them under control, but always solicit your employees' opinions to make them feel that their opinions are valued. This way, anyone who wants to can have some type of input into the company. When your people can do this, it increases their personal stake in the company. They begin to feel like a part of the organization, rather than someone who gets paid to push papers.

MOTIVATING PEOPLE

Remember that one of the most important things affecting your employees' productivity is their attitudes. Have you ever heard the phrase "Attitude rules"? Of course this isn't 100 percent true, but it may very well be 80 percent true. It is important enough that you should give special and continuous consideration to your people's attitudes. You need to spend some time nuturing everybody's attitudes (even your own). This must be continual, as attitudes are not static. Attitudes are subject to change with any change in a situation; continual maintenance is required.

Make your place of business a place where your employees like to go. You have to find the proper balance between productivity and pleasantness. Until now we have talked mainly about productivity and avoiding wasted time, but this isn't the whole story. To make your operations as beneficial as you can, you need to consider your workforce. They are the ones who will have to carry out the company's policies; they must perform the day-to-day functions. In short, without them you will get nothing done and the company will not operate. You need to know what makes your people want to work; what it is that gets them in the mood to work.

Let's start at the beginning. First of all, your people need money to pay for groceries and a place to live. But what else beyond this? Most of your people should have these concerns basically taken care of. At this point their requirements of food and housing are no longer a concern, and are therefore not a source of motivation. Fifty years ago this was all the motivation that many people needed, as it was not so easy to come by, but as times and society change, so do the forces that motivate people. The job here is to find out what will make the difference between an employee who is merely putting in time and an employee who takes pleasure in being good at the job. You need to know what it is that will keep your employees interested and in touch with their jobs.

As we said; the things that motivate people today are quite different from those which moved workers in decades gone by. Now workers are more concerned with feeling that their work is really appreciated, feeling in on things, having interesting work, and having good working conditions. Of course money is still important, but once that hurdle is behind you these are the things that matter to people. These are the things that will matter over the long haul. While these attitudes may change over time, they usually change very slowly, and barring a major economic or national crisis they will probably continue as they are.

In reviewing these facts, it becomes evident that people want a certain amount of pleasure from their jobs. If they don't get this, they are likely to be dissatisfied. Now remember the boss' perspective: He is strongly motivated and has no need for pleasure in the workplace. His concern for the company's doing well is a sufficient amount of motivation. Frequently he will see the little pleasures that his employees want as a total waste of his money. It is true that fringe benefits won't do anything by themselves to make money, but in the real working world they are often necessary. This brings us back to the realization that not everyone thinks the same way you do. The boss doesn't need the little fringe benefits; he desperately wants the company to succeed, and sees these benefits as an unnecessary expense. The employee may see these same benefits as his reason for staying on the job, for working after-hours, or for showing a little more concern than he "has to."

This is not to say that you need to be a psychologist, but that if these are the things you need to do to get results, then you had better start thinking about them. If feeling appreciated is what makes people perform, then your job is to make them feel appreciated.

To complicate this problem, no two individuals are identical. Everyone has his own strengths and weaknesses; individuals each have their own capabilities and desires. You cannot please everybody all the time, but if you never make an effort to figure out what each individual is looking for, you will certainly fail. Your employees' attitudes require continual maintenance; they can either operate like a finely tuned machine, or like a heap. You (as management) make the difference. You can not afford even from a strictly monetary standpoint to ignore this; by ignoring the fact that your people need to keep good working attitudes, you've already made a decision to let them rust into uselessness.

Undesirable as it may be, the fact is that most people want a certain amount of pleasure mixed into their jobs. It's easy enough for the owner to work hard with no pleasure break, but if he expects employees to do the same he will probably lose a lot of them quickly. There is no specific way of interspersing pleasure with work that will work for everyone. The best way of doing this depends on your location, what type of operation you run, the size of the firm, and the individuals themselves.

Ideally, the best motivation that you can give your employees is to instill in them the will to perform; that is, to get them to the point where they work because they want to, not just because they have to. To instill this in your people, the first thing they need to know is: What are our goals? Where are we going? Your employees need something to work toward, so they can feel like they have accomplished something. Next, you need to keep your people working as a team, both among themselves and with management. You need to keep peace in the office, even if it means making some big changes.

As management, you also need to maintain a reputation of being fair. Your people should feel that you are concerned with their feelings. You need to be the leader that they respect. Remember that the best way for you to instill a good attitude in your people is for you to communicate your good attitude to them. Attitudes are contagious, so if yours is legitimately good, your employees will also tend to develop a good outlook on their job.

PLANNING A BUDGET

There are several types of planning that are essential to any electrical-contracting firm. First is overall planning, as was just covered. Next is preparing a budget for the company's operations. This is almost always done on a yearly basis, as virtually everyone's books (and taxes) are set up on a yearly basis.

Along with your yearly budget, here are several other factors that are important to consider:

1. **The general state of affairs for the electrical-construction industry in your area.** What do you expect for the next year? While no one is expected to be exactly perfect in their predictions of the future, try to ascertain as best you can what the market will be like in the next year.
2. **Your overhead and profit percentages.** Depending upon your sales volume, market conditions, and your current overhead, you may need to charge more or less overhead and profit to make your plans work properly.
3. **Your market placement.** What types of work do you do? Would you be better off doing more of one type of work and less of another?

Preparing a yearly budget is important to you for two major reasons:

1. To make sure that you make enough money to meet your expenses, or to adjust your expenses and income for the company's maximum benefit.
2. To provide a basis for checking on your progress throughout the year. Plans must change, as things rarely work out exactly as planned. As you go through the year, you can refer back to your original plan (your budget) and see where you are ahead or behind.

In Table 2–2 is shown a yearly budget for a medium-to-large electrical-contracting company's overhead expenses. Figure 2–1 shows a calculation sheet explaining the company's volume and profit projections.

TABLE 2–2 Yearly Overhead Budget

Owner's salary	$ 41,600
Estimator's salary	26,000
Superintendent's salary	20,800
Bookkeeper's salary	19,500
Receptionist/typist's salary	13,000
Taxes, etc. overhead (Employees)	14,000
Loan Payments (trucks)	9,000
Maintenance (trucks)	5,700
Mortgage on office	11,400
Utility bills	3,400
Telephone bills	4,200
Insurance (auto, liability, etc.)	3,500
Gasoline (trucks)	4,000
Office supplies, postage, etc.	2,400
Licenses, etc.	400
Legal fees	1,000
Accountant's fees	2,700
Plan room fees	400
Publications, training, etc.	2,500
Temporary office help	4,000
New and replacement tools	2,400
Bad debts	4,000
Misc. equipment rental	2,800
Interest on business loan	1,350
Travel and entertainment	2,000
Advertising	3,100
Contingency money	2,400
Total	$207,550

The calculations shown use an assumed figure of 15 percent of sales for overhead, which is a consistent figure for electrical contractors. With 15 percent being an average figure, it can be assumed that whatever the company's overhead budget may be, the company should be able to do a volume that is 6.67 times as great as the overhead budget. (1 ÷ .15 = 6.67; if overhead is 15 percent (.15), then the whole volume will be equal to 1 ÷ .15, which is 6.67, multiplied by the overhead.)

The calculations in Figure 2–1 also show how much profit the company stands to make at different levels of volume and profit margins, which is important to know. At the bottom of Figure 2–1 are markup and yield figures, which show what percentage of sales that overhead and profit will be at certain percentages of markup. Let's take the first one that is shown as an example. The chart shows that cost plus 20 percent will yield 16.67

CALCULATIONS

If yearly overhead is $207,550, and we assume that overhead expenses are 15% of total sales, we will have to do $1,384,358 worth of work. The calculation is done as follows:

Solution If overhead is 15%, then the total volume will be 1 ÷ .15, or 6.67 times the overhead cost. $207,550.00 x 6.67 = $1,384,358.50

If the company can do $1.4 million worth of electrical work at a 20% gross profit, it will make a total profit of $72,450.00, as shown:

$1,400,000	×	20%	[.20]	=	$280,000
$280,000	–	$207,550		=	$72,450

Value of Work (in millions of dollars)	Profit Percentage	Profit or Loss ($)
1.4	18	44,450
1.4	22	100,450
1.3	18	26,450
1.2	18	8,450
1.0	18	27,550
1.5	18	62,450
1.5	20	92,450

Assuming a 15% overhead and the required volume, then we obtain these figures:

Percent of Markup	Percent of Sales to Profit
Cost + 20%	1.67%
Cost + 25%	5.0%
Cost + 30%	8.08%
Cost + 33%	10.0%
Cost + 50%	18.3%

Figure 2–1

percent of the total sales (that's sales, not cost) toward overhead and profit. This is the total of the 1.67 percent shown for profit, and the 15 percent that is stated as overhead. For a sample calculation, let's assume that the cost of a certain job is $10,000, and if we mark it up 20 percent, we add $2000 to the cost, which brings the total price up to $12,000. So, we marked up the job by 20 percent, but the markup was 20 percent of cost. The $2,000 mark-up for overhead and profit was 20 percent of cost, but it is only 16.67 percent of the total sale price (2000 ÷ by 12,000 = .1667, or 16.67 percent).

Table 2–3 goes a step beyond this, showing the gross profit (which includes both overhead and profit) for twelve different percentages of markup. Again, it is important to remember that mark up is a percentage of *cost*, not a percentage of total sales.

The footnote for Table 2–3 is an explanation of corrected percentages. It has long been a widespread practice for contractors to mark up the cost of their jobs by 10 percent for overhead, and 10 percent of that total for profit. The problem with this is that the figures of 10 percent and 10 percent are misleading. The 10 percent figured in for overhead is not 10 percent of sales, as is often assumed; it is only 8.26 percent of sales. So if a contractor has a volume of $400,000 per year, and marks up his jobs the assumed 10 percent to cover his overhead of 40,000 he is going to end up almost $7,000.00 short.

Overhead and Profit Percentages

As you may have noticed, the figures that are shown use a flat percentage for overhead and profit, rather than the 10 + 10, 10 + 7, or 10 + 5 that a lot of contractors have used through the years. This is both more realistic and easier to work with. In real life, almost all electrical contractors have an overhead and profit budget that is about 15 percent of total sales; only the largest contractors are likely to beat this percentage

TABLE 2–3 Markup and Gross Profit Percentages

Markup (%)	Yield (Percent of Sales Toward Gross Profit)
10.	9.09
15.	13.0
20.	16.7
25.	20.0
30.	23.1
33.3	25.
35.	25.9
40.	28.6
45.	31.0
50.	33.3
60.	37.5
66.7	40.

Corrected percentages. A markup of 10 + 10 (10% for overhead, then 10% of that sum toward profit) actually yields 8.26% of sales toward overhead, and 9.1% of sales toward profit. This comes out to a total markup of 21.% above cost.
A markup of 10 + 7 (10% for overhead, then 7% of that sum toward profit) actually yields 8.5% of sales toward overhead, and 6.5% of sales toward profit. This comes out to a total markup of 17.7% above cost.

by very much. This being the case, the 10 + 10 that a lot of contractors use isn't really 10 + 10, but 15 + 2.3 (10 + 10 gives a total markup over cost of 21 percent, which will yield 17.36 percent of sales toward overhead and profit). By using a flat percentage to apply overhead and profit, comparisons of prices are simpler, and figuring the amount of profit that you actually have in the job is far easier.

There is another school of thought in the construction industry. Its view is that a contractor should use different percentages of overhead and profit for material and labor. The argument is that overhead and profit can be more accurately applied this way. While there is nothing wrong with this idea in theory, in actual use it is usually more trouble to implement than it is worth to an electrical contractor. This method is probably useful for some of the other trades (general contractors specifically), but very seldom for electrical contractors. For almost all electrical contractors the relationship of material to labor is about the same for most of their jobs, so the final percentage that is used for markup will be almost identical anyway. With this being the case, spending the time to figure out separate percentages has no real benefit to the contractor.

What really matters in charging overhead and profit is simply that you charge enough. The only way to verify this is by going over your books (usually at the end of the year) and finding out exactly how much you spent for overhead, how much profit you made, and what your total sales were. Again in this application, simplicity and accuracy argue for using a flat percentage in applying overhead and profit.

In the sample overhead budget (Table 2–2) notice that payroll taxes are included for overhead employees only. Only the taxes on these employees should be included in the overhead budget. All payroll taxes and other labor expenses should be included directly in the hourly labor rate. These expenses typically include FICA (Social Security), unemployment taxes, Workman's Compensation Insurance, Union Funds, Contractor's Associations, Apprenticeship Training Funds, and the like. This is calculated as an hourly figure. By adding all of your electrician's hourly rates to these expenses, and dividing this total by the number of electricians, you can calculate your company's average hourly labor cost figure.

Keeping key employees. Obviously, it is important for a company to keep its key employees. Especially dangerous is the threat of a good employee being stolen away by another electrical contractor. The best way to avoid this problem is to give your top employees a yearly bonus based upon a percentage of the company's profits. This idea can be somewhat unpleasant to the owners of many companies. The owner has taken all of the risks, worked the hardest, and sweat more than anyone else. Furthermore, the profits are under his control, and money already in the pocket is difficult for most people to give away. While these misgivings about giving away some of the company's profits are certainly understandable, it is very much to the owner's long-term benefit to do so. Keeping key people while giving away a certain percentage of the company's profits is far better than hanging on to all of the profits while losing the company's most valuable assets.

In Figure 2–2 is shown a plan for the distribution of a company's profits. Right from the start, the first $10,000 goes directly back to the company. This figure will of course vary upon the size of the company, but these funds are to be used for the purchase of new equipment, tools, perhaps a down-payment on a new building, or some other investment for the company's future. Next are bonuses for the key people in the firm, the owner, estimator, superintendent, and bookkeeper. These people and the company will split most of the profits over $10,000 as shown. The remaining 25 percent will go to other employees, probably in the form of a year-end bonus. This works out well for several reasons.

1. Your key people will know that you are serious about them, and that they have a stable future with your company. This will make them a lot less likely to change companies.
2. Any company trying to steal your key people would have to make them the same deal that you have (or at least comparable), which many of them will not be willing to do, especially for someone new to their company.
3. These people's productivity will increase, because they know that they will see a substantial benefit from the company's making a profit.

DISTRIBUTION OF PROFITS

$0.00—$10,000

All profits up to $10,000.00 will go directly back to the company for the purchase of new equipment or similar uses.

$10,000 and OVER

All profits over $10,000.00 shall be distributed as follows:

25%	To the company
20%	To the owner
15%	To the estimator
10%	To the superintendent
5%	To the bookkeeper
25%	To other employees

Figure 2–2

No employee works as hard for someone else as they do for themselves. By giving them a percentage of the profits, their attitude changes from that of an employee to that of a partner, and their productivity will change accordingly.

4. These people will certainly not want to change jobs until they receive their year-end bonuses. By giving these bonuses as the traditional year-end (Christmas) bonuses, your people get their payments at the slowest time of the year. The other contractors who were trying to steal your people in July will probably not be interested in January.

SCHEDULES

Schedules, if properly used, can be one of the best management tools that an electrical contractor can have. By using schedules, you can stay several weeks (or even months) ahead of problem situations. Granted, it can be difficult to get a schedule out of some general contractors, but if you can (usually you can if you try hard enough), you can plan your manpower requirements, material deliveries, and even your cash-flow requirements.

Figure 2–3 shows a sectionalized job schedule. You will notice that it is divided into seven sections, which are for the seven basic sections of electrical construction projects. The sections are as follows.

Section One	Slab, site, set-up
Section Two	Interior raceway
Section Three	Distribution equipment
Section Four	Wire
Section Five	Light fixtures
Section Six	Trim
Section Seven	Other

To use this schedule, the labor hours for each section of the job should be divided by 40 to give the number of worker/weeks which are required for each section. These figures will be compared with the General Contractor's schedule, and filled in at the proper locations in the schedule.

At the bottom of the schedule is a horizontal column marked Total. This column is for the total number of workers that will be needed for each week. To get this figure, add the number of workers shown for each section during that week, and enter this total in the Total column. This will tell you how many workers you will need on the job at any specific time. This way, if you will need to hire any new people, or rearrange some people, you will have plenty of time to do so.

Because of the way schedules change, this schedule has a place at the top to record the date that it was prepared, and by whom. This eliminates the problem of figuring out which schedule is the most recent

Sectional Job Schedule

Job Name ______________________ Date Prepared or Revised ______________________

Job Number ______________________ By ______________________

Month																								
Week																								
Section 1																								
Section 2																								
Section 3																								
Section 4																								
Section 5																								
Section 6																								
Section 7																								
Total																								

Section 1 - Slab, Site, Temp.: ______ hrs. = ____ Worker/Weeks. Section 2 - Interior Raceway: ______ hrs. = ____ Man/Weeks.
Section 3 - Dist. Equipt.: ______ hrs. = ____ Worker/Weeks. Section 4 - Wire: ______ hrs. = ____ Man/Weeks.
Section 5 - Light Fixt.: ______ hrs. = ____ Worker/Weeks. Section 6 - Trim: ______ hrs. = ____ Man/Weeks.
Section 7 - ________________: ______ hrs. = ____ Man/Weeks.

Figure 2–3

one. At the bottom of the schedule is a place for the labor-hour calculations to be performed. Figure 2–4 shows a completed sectionalized job schedule.

Similar to the sectional job schedule is the composite schedule, as shown in Figure 2–5. The purpose of this schedule is to ascertain the total manpower requirements for the whole company at any given time. This is done by taking the number of workers needed on each individual job (week by week), and filling these totals into the proper location on the composite schedule. These columns can then be totaled to give you the total number of workers that the company will need for each week. Figure 2–6 shows a completed composite schedule.

Using schedules is important to you for one main reason: It keeps you in control. By knowing what to expect, you buy yourself some time to deal with the situations that are coming, and thereby save yourself a lot of grief. The unfortunate fact is that most electrical contractors have a somewhat miserable time at their jobs—they are continually trying to fix crisis after crisis. Some of this occurs naturally in electrical contracting, but much of it could be avoided by guiding the company into situations rather than waiting for situations to develop and then reacting with haste.

In Figure 2–7 is shown a unique type of schedule called a *Cash-flow Schedule*. This schedule is a little more complex than the others, and requires a little more work to prepare, but it is well worth the extra effort. Cash flow is certainly one of the more difficult problems that electrical contractors face. This schedule projects your company's cash needs—how much cash you are likely to receive, and when.

The information needed to fill out this schedule comes from either the sectional job schedule or the composite schedule. If you take information from the sectional job schedule, the cash-flow projection will be for the job that the sectional job schedule is based upon. If you take your information from the composite schedule, however, the cash-flow projection will be for the whole company, not just one individual job.

Aside from the information from the other schedules, you will need the following:

1. **Labor $ (labor dollars).** To obtain this figure, multiply the number of worker/weeks by 40 to get the total number of worker hours for each particular week. Then multiply worker/hours by your company's average hourly rate to get labor dollars.
2. **Material/Labor ratio.** This is quite simply material dollars divided by labor dollars. One way to get this figure is to review the previous year's books; see how much was spent on materials, how much was spent on labor, and then divide. You can also get this information from the summary sheets of recent estimates that your company has

Sectional Job Schedule

Job Name: HAUSER CO. — Date Prepared or Revised: 3/7/88

Job Number: 3804 — By: J.M.

Month	APRIL				MAY					JUNE				JULY				AUGUST				SEPT.			
Week	4	11	18	25	1	8	15	22	29	5	12	19	26	3	10	24	31	7	14	21	28	4	11		
Section 1	2	–	1	–	–	–	–	–	–	–	–	–	–	–	–	–	–	–	–	–	–	–	–		
Section 2	–	–	–	–	4	4	4	4	3	2	1	–	–	1	1	–	–	–	1	–	–	–	–		
Section 3	–	–	–	–	–	–	1	2	–	–	1	–	–	1	–	–	–	–	–	–	–	–	–		
Section 4	–	–	–	–	–	–	–	–	–	2	3	3	2	–	1	–	–	–	–	–	–	–	–		
Section 5	–	–	–	–	–	–	–	–	–	–	–	–	–	1	2	4	4	2	2	–	–	–	–		
Section 6	–	–	–	–	–	–	–	–	–	–	–	–	–	–	–	–	–	–	–	3	3	1	1	–	
Section 7	–	–	–	–	–	–	–	–	–	–	–	–	–	–	–	–	–	–	–	1	1	1	–	–	
Total	2	–	1	–	4	4	5	6	3	4	5	3	2	3	4	4	4	2	3	4	4	2	1	—	

Section 1 - Slab, Site, Temp.: 120 hrs. = 3 Worker/Weeks. Section 2 - Interior Raceway: 1,000 hrs. = 25 Man/Weeks.
Section 3 - Dist. Equipt.: 200 hrs. = 5 Worker/Weeks. Section 4 - Wire: 440 hrs. = 11 Man/Weeks.
Section 5 - Light Fixt.: 600 hrs. = 15 Worker/Weeks. Section 6 - Trim: 320 hrs. = 8 Man/Weeks.
Section 7 - FIRE ALARM : 120 hrs. = 3 Man/Weeks.

Figure 2–4

Composite Schedule

Date Prepared ____________________ By ____________________

Month																								
Week																								
Total Number of workers needed																								

Figure 2–5

Composite Schedule

Date Prepared 4/12/87 By J.R.

Month	August					September				October				November					December				Jan.	
Week	3	10	17	24	31	7	14	21	28	5	12	19	26	2	9	16	23	30	7	14	21	28	4	11
Glen-Brook	–	–	2	–	4	4	5	7	7	9	9	9	8	6	5	7	5	5	4	3	4	3	2	3
Valley Products	10	10	9	7	9	7	6	5	5	4	2	1	–	1	–	–	–	–	–	–	–	–	–	–
K Mart	–	–	–	–	–	–	–	–	1	–	2	–	4	9	9	12	10	9	9	5	7	5	6	5
Wilson Co.	4	4	7	7	9	7	9	5	5	4	3	4	2	1	2	1	1	–	–	–	–	–	–	–
No. Side Golf Club	7	5	4	2	4	2	1	–	1	–	–	–	–	–	–	–	–	–	–	–	–	–	–	–
Total Number of workers needed	21	19	22	16	26	20	21	17	19	17	16	14	14	17	16	20	16	14	13	8	11	8	8	8

Figure 2-6

Cash Flow Schedule

Prepared By: ____________________ Date: ____________________

Month																								
Week																								
Labor $																								
Material $																								
Subtotal																								
Overhead																								
Total Cash Needed																								
Money Received																								
Cash Flow + or – (Positive or Negative)																								
Cumulative Cash Flow (Positive or Negative)																								

Ave. Hourly Labor Rate - $________ Mat./Labor Ratio -________ Ave. Overhead %________ Ave. Profit ________ % Ave. Retainage ________

Labor $ = Worker/Weeks × 40 × Ave. Rate   Mat $ = Labor $ × Ratio (Entered 2 weeks after Labor $)

Overhead = Sub-Total × Ave. % of Overhead   Cash Needed = Overhead + Sub-Total

Money Received = Total Cash Needed + Profit – Ave. Retainage (Entered 4 weeks after cash needed)

Figure 2–7

done. Depending upon your location, this figure will vary from around two to four, meaning that material is two to four times greater than labor.

3. **Average Overhead.** Again, this figure is best obtained from your past year's records, but usually averages 15 percent.
4. **Average Profit.** This is the average percentage of profit (above overhead) that you figure into your jobs.
5. **Retainage.** This is the average percentage of retainage (if any) that you expect to be held back.

You will notice that some of these figures are average figures. You will also be using some average figures for time delays (the time from when the material is delivered to when you pay for it, and the time from when the work is done until you get paid for it). Because these are average figures, their accuracy is not going to be perfect, although they will be very close. It is possible to calculate cash flow more accurately, but it is so difficult that it is rarely done. The accuracy that you get by using this schedule will be more than good enough for any practical use. If you need to adjust the time-delay figures from those which are shown, you can certainly do so.

Here are the steps to filling out the cash-flow schedule:

1. From your other schedules, calculate your Labor $ figures, and enter them in the proper places.
2. Multiply the Labor $ figures by the Material/Labor radio. Then enter this figure in the Material $ column two weeks after the Labor $ figure from which it was taken.
3. Add the Labor $ and Material $ figures for each individual week. Enter these sums in the Subtotal column.
4. Multiply the Subtotal for each week by the Average Overhead Percentage. Enter this total in the Overhead column.
5. Add the Subtotal and Overhead figures for each week, and enter these figures in the Total Cash Needed column.
6. Take the Total Cash Needed figures for each week, and add to them the average percentage of profit. For example, if the Average Profit percentage is 7 percent, multiply the Total Cash Needed by 1.07. Then deduct the retainage that you expect to be held from this amount. Enter the resulting figure in the Money Received column four weeks after the Total Cash Needed figure from which it was taken.
7. Find the difference between the amount in the Money Received column and the amount in the Total Cash Needed column. Enter this figure in the Cash Flow column. If the Money Received figure is larger

than the Total Cash Needed, the number will be positive. If the Total Cash Needed figure is greater, the Money Received amount will be negative.

8. Keep a cumulative total of cash flow to date by adding all previous weeks (positive or negative) to the week you are currently figuring. Then enter this figure in the Cumulative Cash Flow column.

A completed cash-flow schedule is shown in Figure 2–8.

By projecting your cash-flow needs, you give yourself time to work out arrangements to cover your cash shortages. This usually prevents the problem of not having enough cash to meet your payroll.

OFFICE MANAGEMENT

Your office is the heart of your company's operations, and therefore it deserves a lot of your attention. The main object is to make your office as conducive to work as possible without becoming a slave driver.

One of the first things that you want for your office is for your employees to work without interruptions. Interruptions are particularly difficult for estimators and bookkeepers, whose jobs require intense, careful thought. After an interruption it can take people quite a while to regain their concentration. Furthermore, of all the mistakes that estimators and bookkeepers make, perhaps three-quarters or more are caused by interruptions.

Try to lay out your office so that employees have as much privacy and quiet as possible. Keep them free from salespeople and phone calls. Phone calls are probably the worst of distractions, as their obnoxious ringing demands attention. Have the ringers on these employees' phones disconnected, and let someone else answer the phones. Don't feel obligated to talk on the phone; the phones are there to serve you, not for you to serve them. Have you ever noticed how most people will talk to someone on the phone sooner than they will talk to a customer who has taken the time and effort to come to their place of business? Have messages taken instead, and return the calls at a convenient time (always return your calls). Try to do all of your phone calling at one time, and get it out of your way. Phone calls are important, but don't let them control you. Take only the most important phone calls, and let the others call back later.

Talk to your suppliers, and tell them the times when you would like their salespeople to call on you. Certainly you will have to make exceptions, but it will make things a lot easier if you set a schedule for them to follow. Let them know that no other times will be acceptable, unless agreed upon in advance. You should also set a time for their phone calls.

Cash Flow Schedule

Prepared By: P. R. Date: 10/12/87

Month	June				July			August				Sept.					Oct.				Nov.		
Week	16	23	30	7	14	21	28	4	11	18	25	1	8	15	22	29	6	13	20	27	3	10	17
Labor $	200.-	500.-	1,000.-	1,000.-	800.-	800.-	600.-	700.-	400.-	200.-	—	—	—	—	—	—	—	—	—	—	—	—	—
Material $	—	—	800.-	2,000.-	4,000.-	4,000.-	3,200.-	3,200.-	2,400.-	2,800.-	1,600.-	800.-	—	—	—	—	—	—	—	—	—	—	—
Subtotal	200.-	500.-	1,800.-	3,000.-	4,800.-	4,800.-	3,800.-	3,900.-	2,800.-	3,000.-	1,600.-	800.-	—	—	—	—	—	—	—	—	—	—	—
Overhead	30	75	270	450	720	720	570	585	420	450	240	120	—	—	—	—	—	—	—	—	—	—	—
Total Cash Needed	230	575	2,070.	3,450.	5,520.	5,520.	4,370	4,085.	3,220.	3,450.	1,840.	920.	—	—	—	—	—	—	—	—	—	—	—
Money Received	—	—	—	—	221.	554.	1,993.	3,322.	5,316.	5,316.	4,208.	3,934.	3,101.	3,322.	1,772.	886.	—	—	—	—	3,772. ← Retainage	—	—
Cash Flow + or − (Positive or Negative)	-230.	-575.	-2070.	-3450.	-5,299.	-4,966.	-2,377.	-763.	+2,096.	+1,866.	+2,368.	+3,014.	+3,101.	+3,322.	+1,772.	+886.	—	—	—	—	+3,772.	—	—
Cumulative Cash Flow (Positive or Negative)	-230.	-805.	-2,875.	-6,325.	-11,624.	-16,590.	-18,967.	-19,730.	-17,634.	-15,768.	-13,400.	-10,386.	-7,285.	-3,963.	-2,191.	-1,305.	—	—	—	—	+2,467.	—	—

Ave. Hourly Labor Rate - $ 12.75 Mat./Labor Ratio - 4.0 Ave. Overhead % 15 Ave. Profit 7 % Ave. Retainage 10%

Labor $ = Worker/Weeks × 40 × Ave. Rate Mat $ = Labor $ × Ratio (Entered 2 weeks after Labor $)

Overhead = Sub-Total × Ave. % of Overhead Cash Needed = Overhead + Sub-Total

Money Received = Total Cash Needed + Profit − Ave. Retainage (Entered 4 weeks after cash needed)

Figure 2–8

Explain to them that you are not trying to be difficult, but that interruptions are a problem for you, and that you want to avoid them. Make exceptions to the schedule as often as is necessary. An added benefit is that if the salesmen are restricted to certain hours they tend to get right down to business, rather than spend a lot of time on gossip and chit-chat.

One particular problem for estimators is with job foremen calling in with questions. This takes the estimator's mind completely off the job that he is working on, and focuses his attention on another job. It tends to confuse the two jobs in his mind, which makes for problems. As with your salespeople, let your foremen call in at either the beginning or end of the day unless the problem absolutely cannot wait.

Your office people need a clear sense of direction within the company, as do all employees. This also applies to their individual jobs; they need to know exactly what is expected of them. You need to clarify each person's responsibilities. Beyond this, they also need to know exactly what they can expect from you. If each of you in management doesn't know exactly what the others are supposed to do, you may find one person expecting another to do something that the other person is sure is not his to do. If your employees don't know what they are supposed to do, things may not get done. This isn't really fair to the employees, as they may be expected to do something (and be in trouble for not doing it) that they never were explicitly told to do. One of the more difficult things to realize in life is that not everybody thinks the same way that you do. What you think of automatically may never even come into someone else's mind.

Lists of Responsibilities

The best way to clarify each person's responsibilities is simply to make lists of them for your office people. This way there are no misunderstandings; your employees know exactly what you expect them to do. Give a list of responsibilities to all of your company's key people, and then go over the lists with them. Make sure that they understand everything that is shown on the lists, and that they have no unanswered questions. Let them know that they are expected to do everything that is shown on the list, without fail.

Tables 2–4 through 2–8 show lists of responsibilities for a typical electrical-contracting company. If you need to modify these lists for your own company, take some time and go over all of the lists together. Itemize each separate activity that must be done in your company, then cross-check each activity with your lists of responsibilities to make sure that each activity has been assigned to a certain individual. The important thing is that you have no gaps. Remember that, as management, it is your responsibility to make sure that nothing is left undone.

Once you have your lists of responsibilities settled, be sure to enforce them. Hold your people to the rules. If you don't enforce these rules,

TABLE 2–4 List of Responsibilities—President or Owner

1. Prepare the company's long-term planning.
2. Prepare the company's yearly plan.
3. Review the company's progress on a monthly basis.
4. Verify that every major activity is assigned to a particular individual.
5. Make sure that each individual meets their responsibilities.
6. Review each major job on a weekly basis.
7. Keep everybody in the company happy and working.
8. Stay in touch with our customers, and see to it that they are satisfied with us.
9. Stay current with market, financial, and legal trends.
10. Meet with the accountant once per month.
11. Meet with the attorney as required, but at least once per year.
12. Meet with insurance salesperson as required, but at least several times per year.
13. Keep the office clean and organized.
14. Fill in for any worker who is off, except if someone else is responsible for these duties.
15. Make a concentrated effort to maintain a good attitude in all of our people.
16. Assist all employees.

TABLE 2–5 List of Responsibilities—Estimator

1. Prepare estimates and bid new jobs.
2. Review jobs in progress with the superintendent weekly.
3. Review and verify weekly all invoices for jobs you estimated.
4. Review all job accountings weekly.
5. Prepare and execute all POs for new jobs.
6. Stay up-to-date with new products and market trends.
7. Prepare submittals for new jobs; verify their approval.
8. Work with the superintendent and bookeeper to prepare monthly draw requests or other billings.
9. Assist superintendent in preparing change orders or other estimates.
10. Assist president/owner in future planning.
11. Keep the office clean and organized.
12. Assist coworkers as time permits.
13. Teach apprenticeship classes as required.

you're company will be chaotic. These lists of responsibilities are your company's rule book. Your people should know that their actions are expected to conform to the rules, and if they don't there will be problems.

As you make up these lists, assign a back-up for each important operation. This way, if one person is sick or on vacation, there will be someone available to fill the gap; and because they know that they will be required to fill in if the other person can't be there, the back-up will usually do a very acceptable job. This back-up responsibility is often given to the company owner or president. This works well because the owner or president usually knows how to do almost everybody's job reasonably well, and if he doesn't know, he will usually care enough to make sure that the job gets done one way or another.

TABLE 2–6 List of Responsibilities—Superintendent

1. Visit each job site frequently, preferably daily.
2. Discuss each job in detail with the job's foreman at least once weekly.
3. Discuss each job in detail with the GC or owner at least once weekly.
4. Prepare and handle all change orders.
5. Review each job with the estimator who figured them weekly.
6. Maintain a good relationship with the other major trades on all jobs.
7. Take care to maintain a good attitude among our foremen and workers.
8. Prepare POs for any material needed once the job is in progress.
9. Order inspections as required.
10. Take care of all correspondence regarding these jobs.
11. Maintain good relationship with any job architects or engineers.
12. Maintain a good relationship with the local inspectors.
13. Prepare estimates for small jobs as required.
14. Keep the office and your truck clean and organized.
15. Teach apprenticeship classes as required.

TABLE 2–7 List of Responsibilities—Bookkeeper

1. Prepare weekly payroll.
2. Prepare payables.
3. Prepare billings.
4. Prepare all job-accounting forms.
5. Check all POs and job costs when the estimator is off.
6. Maintain an accurate, up-to-date set of books.
7. Assist the accountant in preparing financial statements.
8. Keep the office clean and organized.
9. Help coworkers as time permits

TABLE 2–8 List of Responsibilities—Secretary/Receptionist

1. Take care of all the mail, both incoming and outgoing.
2. Take care of all incoming calls.
3. Prepare all correspondences (typing).
4. Assist estimator in tabulating estimates and taking phone quotes.
5. Maintain the company's filing system in good order.
6. Keep the office clean and organized.
7. Assist coworkers.
8. Take care of all interoffice messages.
9. Assist bookkeeper, and help take care of the bookkeeper's responsibilities if the bookkeeper is off.

Scheduling

The next step in office management is scheduling your employees' activities. This works well for several reasons:

1. Schedules activate people; they put employees in gear and get them in motion. It is quite natural for many people to sit around and take a break until something pressing comes along, or until they get

bored. You don't want to be a slave driver, but your employees will be just as happy (usually happier) accomplishing something as they will be doing nothing. Give them a reason to move.

2. By scheduling all of your work, you can make your employees work together. This eliminates a lot of the confusion that is found in most offices. Route your work so that what one person needs to have done before he can do his job is scheduled to be done before he needs it. Besides making your internal operations a lot smoother, you will eliminate a lot of the *Joe is mad at Jim because Jim didn't have the invoices ready* types of problems.
3. By using schedules you have a way of checking on people's progress. You can see if they've gotten the work done that they were supposed to get done that morning. You give yourself a standard to compare your employees' deeds to. If there is something that is a problem, or that doesn't get done, you can know almost immediately. Without having a system of checking your people's actions, you may not find out about problems until you have an irate customer on the phone talking about a lawsuit. It is management's job to make sure that everything that is supposed to get done actually gets done; using schedules gives management a way of making sure before it hurts.

The only problem with using schedules is that some people may resent them, although this need not cause you much of a problem if you handle it correctly. The problem of resentment is not really caused by the schedules, but rather by the perception of the individual that they are becoming merely a number, or a robot; that they, as individuals, are not respected or cared for. If you present your decision to use scheduling as being for the good of the company's operations, and not for putting your people in chains you can avoid this problem very effectively. Figures 2–9 through 2–12 show schedules for a typical electrical-contracting company.

A clipboard system is another technique that works quite well for making sure that things get done. Put up a clipboard for each person next to their desks. Then, if anyone has a message for that person, they can write it down and place it on the clipboard; this goes a long way in making sure that interoffice communications are received. Even if messages can be given verbally, have your employees write them down. Using verbal messages is asking for misunderstandings and forgotten details.

The beauty of the clipboard system is that you can use it to automatically assure that what is supposed to get done actually gets done. You can do this by making one hard and fast rule: All clipboards will be left empty at the end of each and every day. By doing this, you assure that all critical messages are acted upon daily. This is very important to any contracting firm, with the multitude of details and changes that must be taken care of in the course of any construction project. Forgotten details

SCHEDULE
PRESIDENT OR OWNER

Monday—Friday

A.M.	Assist superintendent or others in getting the day started. See to it that everybody is working and happy. Other activities.
P.M.	Make sure that everyone is getting their work done. Other activities.

Saturday

A.M.	Review all jobs. General review and planning.
P.M.	As required.

Figure 2-9

SCHEDULE
ESTIMATOR

Monday—Thursday

A.M.	Plan the day's work. Assist superintendent or others as required to get them started. See salesmen, etc. from 8:30 to 9:30 Prepare estimates, bid jobs, plan jobs, etc.
P.M.	Prepare estimates, etc. Talk to salespeople, superintendent, etc. as required from 4:00 until end of the day. Take care of any messages on clipboard.

Friday

A.M.	Review all jobs with superintendent. Check invoices, POs, job accountings, etc. Prepare billings, change orders, etc. Return approved invoices to bookkeeper for payment. Return job folders to bookkeeper. Give billing information to bookkeeper or secretary for preparation.
P.M.	Prepare estimates, orders, see salesmen, etc. as required. Take care of any messages on clipboard.

Saturday

As required.

Figure 2-10

SCHEDULE
SUPERINTENDENT

Monday—Thursday

A.M.	Get everyone going, work out any problems with scheduling or materials. Superintendent should be the first person at the office in the morning, preferably an hour before anyone else. Visit job sites, coordinate with foremen, GCs, inspectors, architects or engineers, etc.
P.M.	Prepare any change orders, correspondence, orders, schedules, billings, estimates, etc. Visit job sites, or other activities as required.

Friday

A.M.	Review all jobs with estimator. Check invoices, POs, job accountings, etc. Prepare change orders, billings, etc.
P.M.	Visit job sites, or other activities as required.

Saturday

As required.

Figure 2–11

SCHEDULE
BOOKKEEPER

Monday

Make weekly material update to job cards.

Tuesday

Prepare accounts payable and issue checks.

Wednesday

Prepare weekly payroll.

Thursday

Make weekly labor update to job accountings. Prepare invoices and job folders.

Friday

Prepare all billings.

Take care of collections.

Figure 2–12

and lack of coordination cost construction companies money. Sure, human error still comes into the picture, but you will never be able to do away with that totally. By making sure that all communications are acted upon, you are effectively doing away with any problem other than the failure of your people to do the tasks that are plainly recorded in their responsibility lists. If you will make the effort to follow this system, your office (and entire company) will operate better than 98 percent of your competition's.

Make sure that everyone in your office (the foremen who run your projects too) have a calendar. The type of calendar isn't important; the important thing is that everyone has a specific place to write down notes that will be seen at a future date. As with the clipboard system, using calendars is a critical factor in making sure that everything gets done. Some people would prefer to use an appointment book, some desk calendars, and some wall calendars. Whatever it is that they like, make sure that they use it and refer to it every day without fail.

Acting or reacting. To many contractors, the type of set-up that is described here may sound rather odd. The fact is, though, that orienting your company this way will make things easier for all involved, whether boss or employee. One reason that it sounds very different is this: Conventional contracting companies tend to react to situations only after they develop into something that has a significant impact, then the situations are handled hastily. The type of set-up explained here is designed to act upon situations before they become a problem. Sometimes this system can even help anticipate a problem and correct it before it can cause any trouble. By anticipating upcoming situations, your company can handle them in an orderly, efficient manner. This also avoids the problems of lost good will and a tarnished reputation, along with a host of other problems.

One other benefit (and perhaps more important) of this type of system goes directly to the company's management; to the owner in particular. The real benefit is peace of mind. With an electrical contracting company operating in the traditional manner, an earth-shattering crisis is a typical occurrence. By using this system it should become a lot more uncommon. You won't have to worry about things being forgotten. Checking up on your employees becomes easy and painless. Almost everything that your company does will be handled more easily. Sure, it will take some effort to set up this type of system, but the savings of mental anguish alone are reason enough to do it. The average electrical-contracting company that changes over to this type of system can expect to increase the efficiency of their operations by 15 to 20 percent. That's not bad for an investment of a little time and effort.

MEETINGS

As mentioned earlier, frequent company meetings can help your people work together. Here is a typical meeting schedule for an average-sized electrical-contracting company.

Office meeting. Once per month, on the first Thursday. Typically, this is a dinner meeting at a local restaurant, with the company picking up the tab. Take care to keep the conversation centered around the office.

Journeyman's Meeting. Once per month, on the third Wednesday. This is a general rap-session for all the electricians. The apprentices who meet every Wednesday night are also included. When available, one of the local suppliers is invited to give a demonstration, preferably of a new product or technique. Some refreshments are served. This meeting typically runs from 6:30 P.M. to 9:30 P.M., at the office.

Apprentice Training. This meeting is held every Wednesday evening from 6:30 to 9:30 at the office. Typically the owner, estimator, or superintendent trade-off teaching shifts of several weeks in a row. No refreshments. Attendance is required for apprentices.

OFFICE LAYOUT

For your office's physical arrangement, the critical factors are these:

1. **Your workplace must be conducive to working.** Interruptions are to be avoided, even at great cost. Keep people far enough apart so that they won't bother each other or get into too much conversation. Make your office well lit, quiet, and comfortable.
2. **Keep your office organized.** Be a stickler on this. Have a place for everything, and keep everything in its place. This will certainly save you a lot of time searching for lost papers, etc.; but the really important thing is the professional attitude that it will convey to your people. By requiring that your office stay clean and organized, you convey the attitude that "We are professionals here, we are serious about this company, and we aren't playing any games." Believe it or not, how you keep your office makes a very loud statement to your employees; they will notice, and follow your lead.
3. **Keep your warehouse clean, well organized, and completely stocked.** Depending upon what type of operation you run, you might want to have almost as much material as a wholesale house in stock, or almost nothing at all. Whatever amount of stock it is that you need,

make sure it is there when you need it. What was said about keeping your office organized applies to your warehouse also, with one difference: Now it will be your electricians who will hear the statement you are making about your company. They really do notice, whether consciously or unconsciously.

3

JOB MANAGEMENT

By their very nature, electrical contractors get their work job-by-job; that is, all of their activities and operations are grouped together in jobs, one job at a time. So it is essential to any profitable electrical contracting company that they manage their jobs well. Managing jobs either well or poorly makes the difference between profit and loss.

JOB PLANNING

The sad fact is that most electrical contractors do only a hit-and-miss job of planning their projects, which is tantamount to throwing money away. What happens to these contractors is that they get in a big hurry, and do only the work that appears to be unavoidable. Because job planning can be put off for a period of time, it usually does get put off. Then it is usually forgotten, and ends up falling in the foreman's lap, thus giving the foreman a lot more work than he has time to do. Is it any wonder that so many jobs just barely break even? First there is no good job planning, and then the most important man on the job is given a great overload before the job even starts.

The main problem is that the average electrical contractor only reacts to problems, rather than acting before a problem becomes unavoidable. While almost anyone in the construction industry would agree that acting on a situation before it becomes a problem is the right way to do things, there is a catch: It requires discipline. It requires that you do something now, even though you could put it off with no immediate harmful side effects.

But there are bad side effects, it's just that they take a few weeks or months to show up. When a project is not planned well, these are the dangerous situations that typically occur:

1. Rather than controlling the course that the job should follow, you let the job chart its own course, like letting a sailboat go by itself in the wind. Some things can be corrected midstream, but there are many things that can never be corrected once the job is under way; their benefits are lost forever.
2. By putting off planning, you avoid a certain amount of work, but you will have to do a much larger amount of work as the job progresses. It would be a lot cheaper to do this work in a quiet office than on a hectic jobsite. Also, putting this load on the job foreman will hurt his attitude (and all of the workers' attitudes through him), and prob-

ably cause him to make a lot of mistakes, just because of his hurry to get everything done.

3. By not taking care of things in advance, but only reacting when something is more important than anything else that needs to be done, you will make a lot of people unhappy with you. This is a big part of why so many people have bad feelings about the construction industry. Some contractors hardly do anything unless someone is threatening them; they create hard feelings wherever they go.

Job planning is often the most important step in job management. If you take the time to plan your jobs well, the rest of your job management will be a lot easier to implement, and a lot easier to recognize as working well or not.

Your pre-job planning should be done as soon as you know for sure that you will be doing the job (usually this means as soon as you get the contract in your hands). Set aside a certain amount of time—an hour is usually enough for a small job, and perhaps a day for a large job—and get the estimator who figured the job, the superintendent, and perhaps the company president all together. Then decipher the job from beginning to end. You want to consider every facet of doing the job. This includes not only the actual electrical installation, but also the business arrangements, working with the architect or engineer, and working with the other contractors who will be working on the job.

A list of things to consider in pre-job planning is shown in Table 3–1. Depending on what types of jobs you do, there may be things shown in this list that don't apply to you, or there may be some things that you will need to consider that are not shown in this list. This is a basic list of topics that need to be covered in pre-job planning.

TABLE 3–1 Pre-job Planning

When are we scheduled to start?
When are we supposed to finish?
How is the GC to work with?
Who are the other major subcontractors? How are they to work with? Are there any services that we can trade with them? (You let us use your backhoe, we get power into your trailer for free, etc.)
Who is the architect and/or engineer? How are they to work with? What are their peculiarities?
Do we expect many change orders?
What are the temporary power requirements?
Will we need to use any specialty subcontractors?
When do we need to have submittals ready?
Will we be allowed to make substitutions?
Who are we going to buy light fixtures from? What about switch gear? How about pipe and wire?
Can we get paid for material stored on the site?

TABLE 3–1 Pre-job Planning

How much will insurance cost for this material and trailer?
Will we rent a storage trailer, or do we have our own?
What does the job schedule look like? Are there any obvious problem areas in the schedule?
Will we have any problem getting deliveries at the required times?
Will we have any problems having enough workers at any particular times?
Will we have any problems with cash flow?
How will payments be processed?
What is the percentage of retainage? Will this percentage change when the job is 50% complete?
How will change orders be handled?
Are we locked in to any certain percentages of overhead and profit? If so, what are they?
Does this GC usually pay on time? If not, what can we do to make him pay us on time?
Do we need to write any letters to inform the GC or architect of problem areas?
Who will be the inspector on this job? What will we have to do to keep him happy?
Will we need any special equipment? If so, will we need to buy it, or can we rent it?
What materials will we need on the job first?
Is there any storage available, or that will be available later in the building?
Which of the job foremen would be best suited for this particular project? Will he be available, or do we need choose someone else?
Are there any particular electricians that would do especially well on this job?
How good are the blueprints?
Should we make any of our own drawings to help the electricians? Will they need details drawn for any specific areas or rooms?
Can we redesign the electrical system at all?
Can we combine two or three circuits in one raceway?
Are there any inadequacies in the plans or specs?
Who supplies the motor starters?
Who does the control wiring?
Who provides disconnect switches for mechanical units?
Who is responsible for pitch pans and sealing roof penetrations?
Are there are particular items or assemblies that we could preassemble in our shop?
Are there any new techniques or tools that would work well on this job?

For the purposes of planning, forget any preconceived ideas that you may have about this job. Then you will have a fresh viewpoint. Treat the job as if it were the first time in your life that you had ever heard anything about it. Let your attitude be: Here is a job that someone has given us. They are allotting us *X* number of dollars to complete it. How can we arrange the job so that it will make us the most possible money?

Take your pre-job planning seriously; it can make a major difference in the outcome of your jobs. Here is a true story that illustrates the point very well:

> A medium-sized electrical-contracting company had given a quote on a small (20,000 square-foot) warehouse. The general contractor that they figured it

> for was a local company that they dealt with often. The job was basically a design/build project, although it was to be approved by a certain architect. Cost was a big factor in the owner's mind, so the electrical system was designed to be very simple and basic. An estimate was made, and the plans were submitted for approval. A couple of months later the general contractor called back, saying that the plans had been approved, the contract had been signed, and that he was ready to get to work.
>
> At this point the electrical contractor's people really went to work. They took the original electrical plans and carefully looked over the raceway layout and circuitry. Then they actually drew up a new set of plans, which had a more economical layout. A new estimate was made from the new plans. Next, they went over the new list of materials and got quotes on the entire list from two or three local suppliers. The purchase order went to one of the supply houses with whom they dealt frequently and had a good relationship with. All of the materials were delivered to the jobsite at one time, with almost nothing left out.
>
> The estimator also drew up several details for the electricians' use. Typically, these were of fixture mountings, or some other crowded or difficult part of the job. A set of working drawings were made up in great detail, even showing the individual runs of conduit detailed as to how many hots, neutrals, switchlegs, and grounds were in each conduit.

Now this may sound like a lot of work, but the final result was this: The company ended up with a total profit of over 40 percent, while doing the job for only about 35¢ per square foot.

Instructing the Workers

A number of years ago a noted estimator was quoted as saying that if you could do a good enough job of instructing your workers, you could cut your labor expenses by 40 percent. The example of the warehouse job tends to prove him correct. The better the instructions that you give your workers, the better your production will be. The instructions that you give your electricians make the difference. So if you don't have a good set of plans, get some. If you can't get some, make your own. If you don't have the time to make your own, you don't have enough time to do the job.

This may come across as a strong statement, but your men can't put a job together unless they know what they are supposed to do. Of course the argument can be made that "We employ journeyman electricians, they know how to put a job together as well as anybody—why should we bother with drawing up detailed plans?" While the argument is understandable (there are many journeyman electricians who are true professionals), the answer is simple: Because it will make you money. Having seen it both ways, I would not ever do a significant job without a detailed set of drawings, period.

You can compare this to putting together two jigsaw puzzles, both of them having about 500 pieces, and both of them perfectly round. One puzzle's pieces are completely blank, all the same color. The other puzzle's pieces are colored so that when the puzzle is complete, it will look like a beach ball. You know which one will be easier to put together.

Redesign. A very important step in job planning and job management is the redesign of various sections of the project. Typically this includes redesigning the job's branch circuitry, and sometimes the light fixtures—whatever can be done to make the job less expensive to do.

By redesigning the projects that you do, you can save quite a bit of money, but there are several things that you need to remember:

1. You may be limited by your contract as to how much redesign you can do. This is especially true of some very large or very closely controlled projects. You will have to read your specs carefully to determine how much redesign you can or can't do.
2. You can't go overboard and try to put nine #12 wires in a half-inch conduit, especially with all of them loaded to capacity. Your work should almost always exceed code by quite a bit. Good workmanship needs to be your trademark if you want repeat business. On the other hand, many engineers design jobs with each circuit in its own raceway, which is not necessary. There is nothing wrong with combining two or three branch circuits in one raceway, provided of course that they are all from different legs (phases) of the system, and are fairly well balanced. By doing this, you save about half of the home runs, which comes out to quite a bit of savings in both material and labor.
3. If you do redesign, you will have to revise the working drawings. All of the changes must be noted so that your workers will know exactly what to do. Don't just explain your plan to your foreman and then send him out to the job to get it done. You will overload him with the job of telling each electrician exactly how to run each circuit.
4. After you redesign a job you should do another take-off, for two reasons. First, so that you obtain an accurate bill of materials to work from when purchasing the materials for the project. Second, so that you will be able to accurately estimate the number of workers hours that will be needed to complete the project. Since labor hours are derived from the types and quantities of the materials that will go into the job, labor figures which are derived from an incorrect material list will be incorrect also.

The easiest type of project to do redesign on is typically design/build, because the nature of the relationship gives the electrical contractor a lot of freedom in designing the installation. But don't ever let yourself get too involved in cost cutting and forget that you are being trusted to provide a high-quality service. In design/build more than in any other type of work, your reputation and trustworthiness are your main selling points. If you lost them you are in trouble.

While redesign is usually easiest to do on design/build projects, it is probably most important for contract-bid jobs. Because of the way that so many electrical contractors cut their percentages of overhead and profit on contract-bid jobs, it can be very difficult to make a reasonable profit on them in many parts of the country. If your market is like this (and almost all are) the only substantial profits you are likely to make are by doing the jobs more cheaply than what you originally estimated. In this situation, doing a good job of redesign becomes crucial.

Preassembly. One technique that has saved contractors a lot of money is that of pre-assembling certain items or groups of items before they are installed. Often the various items are assembled at the company's shop, before they are delivered to the jobsite. How much of this you can do depends upon the individual project. Many contractors used to do this for fixture whips, although some of the fixture manufacturers have started to do this themselves recently. Some of the most common things to preassemble are certain configurations of junction boxes with fixture whips, pre-bending conduits to certain sizes which will be used over and over again during the course of the project, light fixture and mounting assemblies, and some types of distribution equipment.

As you review your new job, keep your eyes open to see if there are any particular parts of the job that would be particularly well suited to preassembly. If you can do some pre-assembly it will help your productivity quite a bit, as you will be doing the work assembly-line style, rather than one-worker-does-all.

PURCHASING

Because materials account for 50–70 percent of most electrical projects, purchasing is a very important function. If you have a good take-off of the job for which you want to purchase materials (and you always should), the actual process of purchasing the material should not be very difficult. After you have done your pre-planning and redesign you should have a very clear idea of how you are going to put this job together. You should also have a very accurate material list. The easiest and best way of purchasing materials for most projects is as follows:

1. Verify that your material list is correct.
2. Make photocopies of your estimate sheets, but cover up or white-out your material price and labor figures. This gives you a plain material list. (You don't want your bid pricing or labor figures out for publication.)
3. Break down your material list into sections so that you can write a separate purchase order for each section of the project. This makes job accounting and job tracking significantly easier. (We will go over section designations and the like in greater detail later.)
4. Give copies of the material list(s) to your local suppliers. Have them bid each section separately.
5. Issue purchase orders to the low bidder(s), provided they can give you the proper materials and deliver them on time.

One question that a lot of growing companies ask themselves is "Should we hire a separate purchasing agent?" If the company is very large ($10 million of volume per year or more), the answer is yes. But for small, medium, and medium-to-large companies the answer is usually no. For those types of companies, it is almost always better for the estimator who figured the job to do the purchasing. No one knows the job better than he does. Not only does he know what types of materials were specified and how serious the engineer is about the spec; but he knows what parts are going to fit together with other parts. Thus the estimator is the only one who can make an intelligent decision as to what parts should be added or deleted if a change is made.

In a company with a purchasing agent (PA), the first step in the purchasing process is for the PA to get all of the information regarding the job out of the estimator's head, and into his own. Obviously this isn't done easily, taking up the time of both men. On the other hand, if the estimator were to do all of the purchasing himself this difficult step would be completely avoided.

The argument for having a PA is that the PA does nothing but purchase material all day long, and therefore does a better job of it than someone who only does it part-time. Obviously there is some truth to this, which is why having a separate PA works well for very large contractors. But unless the company is very large, it would probably be better off having the estimator do the purchasing. The estimator is in close contact with the suppliers; he knows about new materials, techniques, and tools and can stay abreast of the local market. By staying in touch with the current pricing levels, he can make good decisions as to different methods to use in constructing an electrical project.

Figures 3–1 through 3–3 show sample purchase orders (POs). Writing a strong purchase order is an important skill. Electrical contracting (and

(Company Name) **Purchase Order** #

To Demco Supply, Inc. Date 5/7/87

Ship To Wilson Plaza, 2701 SE 43rd Terr., Wilson, UT

Shipping Instructions Deliver as directed.

Deliver The Following —— As Specified Herein

Deliver as requested the material listed on the attached material list "A". Regardless of quantity, prices shall be those unit prices shown in material list "A". These prices are guaranteed for a period of 90 (ninety) days from the date of this order. Any price changes after that time can be made only after we have been given a ten-day written notice. All material shall be listed by Underwriter's Laboratories, and shall conform to the plans and specifications for this project as prepared by Commercial Design Associates, Inc. The above-mentioned plans and specs are available for your review at our office any weekday during regular business hours.

In addition to the aforementioned material list "A", the following items and their unit prices shall be a part of this order:

#12 THHN Cu wire, (solid)	$27.50/M
#10 THHN Cu wire, (solid)	44.00/M
#8 THHN Cu wire, (stranded)	84.80/M

Submit Two Invoices
bearing this PO number

0000 | 5702 | 4 (Company Name) By ____

Received and acknowledged by ____ Date ____

Figure 3–1

all types of contracting for that matter) is a very turbulent business with high stakes. Having a well written purchase order can save you a lot of problems in certain circumstances. While the samples in Figures 3–1, 3–2, and 3–3 don't cover every possible situation, they can give you an insight into how a good purchase order is written. Notice also in the samples

(Company Name) #

Purchase Order

To Mid-State Supply, Inc. Date 11/4/87

Ship To Fort Butler Supply Warehouse, Johnston, W. Virginia

Shipping Instructions Deliveries accepted 7:30 am — 4:00 pm. Call 24 hrs. prior to delivery.

Deliver The Following ——— As Specified Herein

One lot of lighting fixtures, per your quotation dated 11/1/87. This purchase order is contingent upon approval from the engineering firm overseeing this project.
Seven complete sets of submittals shall be delivered to this office no later than 11/11/87. This office shall be advised of scheduled delivery dates no later than two weeks after approved submittals are delivered. All materials shall be furnished exactly as per approved submittals; no changes may be made without our written approval.

Lot Price: $12,015.00 (Twelve Thousand Fifteen Dollars)

Submit Two Invoices
bearing this PO number

1254	8761	5

(Company Name) By ____________

Received and acknowledged by ____________ Date ____________

Figure 3–2

shown here that at the bottom of the purchase order is a space provided for the supplier to acknowledge its receipt; this provides legal evidence that the terms of the order are agreed to by the supplier.

The easiest way to write a PO is usually to write it for "The attached

(Company Name) #

Purchase Order

To Demco Supply, Inc. Date 4/12/87

Ship To Kaiser Factory, 2400 SW College Rd., James, Utah

Shipping Instructions Call job-site 24 hrs. prior. (201) 741-9420

Deliver The Following —— As Specified Herein

One lot of electrical switch gear as per quote #24741-JUT by the Square D company, dated 3/1/87. This material quote is guaranteed by both you and the Square D company to be all of the switch gear required for this project, according to the plans and specifications prepared by Curtis & Blakefield Architects, P.A. All material is guaranteed to meet plans and specs, and is suitable for it's intended use. Material shall be delivered to the job site no later than 4 weeks after you receive approved submittals. 11 (eleven) full sets of submittals shall be delivered to our office no later than 4/28/87. Any delays or defects in the equipment shall be the responsibility of Demco Supply, Inc., who shall be liable for any loss, damage, or expense to us caued by Demco's failure to comply with plans, specs, and/or this order; this includes consequential damages and liquidated damages.

Lot price of above: $29,400.00 (Twenty nine thousand four hundred dollars).

Submit Two Invoices
bearing this PO number

0000 4701 3 (Company Name) By ____________

Received and acknowledged by ____________ Date ____________

Figure 3-3

material list A," as is shown in Figure 3-1. Once you have decided on your suppliers and written your purchase orders, you will need to go over several things with each of your suppliers:

1. **Getting submittals.** This is very important, especially for contract-bid projects. Incorrect or late submittals can cause expensive delays and a lot of hard feelings. If your suppliers make a mess of your submittals, you will be starting the job on the wrong foot. This is your first obligation on the job, so if you do a poor job (whether it's your fault doesn't really matter to the general contractor or the engineer) people will tend to mistrust you from then on.
2. **Delivery schedules.** This can be especially hard for light fixtures and distribution equipment (switchgear). Make sure that you will have the material on the job when you want it. In the real world this means that you may have to push your salesperson a little bit. Make sure your salesperson calls the factory to verify the delivery dates. If possible, get a written confirmation from the factory. (The purchase order shown in Figure 3–4 is specifically written to put the burden of on-time delivery upon the supplier.) Make sure you follow up with the supplier; you don't want the hassles of late deliveries and lawsuits. It would help with large or difficult orders to have the salesperson call the factory periodically to make sure that the order is progressing according to schedule.
3. **Avoiding back-orders.** Back-orders can hurt you more than you may realize. They cause your foremen to shift their workers from one task to another, which cuts productivity dramatically. Come to some type of understanding with your suppliers as to how you will handle back-orders. As every job has its own set of circumstances, there is no one way of handling back-orders that will be correct for every project. This requires you to discuss the matter with your suppliers frequently.

Another facet of purchasing is that of purchasing additional material once the job is in progress. There are almost always items that will be needed that the estimator just couldn't plan on. This is another question to which there is no pat answer, although you should always use a purchase order so that your job accounting won't be thrown off. Some companies like for all purchases to be done through the office, which helps to avoid problems with theft. Also, the office people can often get better prices than a man calling in from a job trailer. On the other hand, some contractors prefer to give the foreman a handful of purchase orders and tell him "If you need something, just go ahead and get it, but be sure to use a purchase order." This method can save time for both the foreman and office; but it is critical that the foreman is trustworthy.

MATERIAL HANDLING

The next thing to review is how you handle your material. The most typical way of handling material for a sizable job is to use a storage trailer located

Daily Diary

Name ______________________ Job Name ______________________

Date ______________________ Job Number ______________________

Men Working ______________________ Weather ______________________

Deliveries ______________________ Inspections ______________________

Comments ______________________

Figure 3–4

at the site. If you do use a trailer, check to see what kind of insurance you will need on its contents. Also check on the location of the trailer. This is a key point as far as productivity is concerned. If your trailer is not located conveniently to the jobsite, your men will end up wasting a lot of time running back and forth to the trailer. Talk to the GC (general contractor) or owner, and make sure you get a good location for your trailer. If the job is an existing building, try to get the use of a lockable room to use for storage. On some jobsites it might be in your best interests to have a storage shed built.

Check the specifications for your job to see if you can get paid for material stored on the site. If you can (commonly you can), you are usually better off having all of your material shipped to the jobsite at the beginning of the project, and then storing it suitably. This benefits you in two important ways:

1. By purchasing your material at the start of the project you avoid any price increases that may come at a later date. You may also get a better price by ordering a large amount of material at one time.
2. You will get paid for all of your material on your first draw. Provided you mark your material up more than the percentage of retainage (and you should), you will generate a good amount of extra cash. This is one way of "front-loading" the job. When you can do this it will help your cash flow quite a bit.

Depending on how well (or how poorly) you handle your material, you can increase or decrease your productivity by a substantial margin. Of course material handling on the jobsite will be your foreman's responsibility, but it is well worth your time to go over things with him to make sure that you agree on how to handle each particular project.

It all comes down to logistics (for those of you who are not familiar with the term, logistics is the military science of supplying and directing troops on the battlefield). Maybe you have never thought about it, but logistics have probably decided the outcome of more battles and wars than any other factor (soldiers who can't get food or ammunition can't fight very well, can they?) Look at your jobs this way and handle your supplies in an orderly, efficient manner. If there is a specific way that you would like to handle material on a certain job, make sure that you explain it well to your foreman.

Ideally, material handling should work something like this: Each day, the job foreman should make a list of the material needed for the next day. This list should be detailed as to what materials are needed at what locations. The next morning, the first order of business should be to have the less-skilled workers (apprentices, helpers) start moving the material around the jobsite as directed. As they deliver these materials to the workers at these locations, they should ask them if there is anything else that they need. If there is, the apprentice should write it down on a list, and bring it to them after distributing the first batch of materials. Doing this keeps the workers busy producing, not spending hours walking back and forth to the trailer and getting distracted along the way.

Tools

Tools, with all of the savings in time that they offer, are very important to an electrical contracting company, but keeping track of them has been a problem for electrical contractors since the industry's beginnings. Here are several ideas that can help:

1. Have some type of organized system of keeping track of your tools. This can be as simple as a card file, or as complex as a computer. The specific method doesn't really matter, just as long as you can refer to it and know where each tool is at any time.
2. Use your system. This is really where the problem lies. The problem is that people get lax about doing their bookwork. Here is a typical scenario: The job foreman walks in the shop at 5:00 in the afternoon after a tough day. It's 95 degrees outside and he's lugging in a power wire-puller. He brings it into the warehouse and starts looking around for the file and can't find it. He's hot and tired, and says to himself "Forget it, it's just not worth it, I've had enough for one day," and leaves the power puller laying on its side in the corner of the warehouse. There are probably about a hundred variations to this story, with similar results. While you can be sympathetic to the foreman, after this happens a couple of times your tool-tracking system is shot. Don't let your people get away with this; enforce your rules.
3. As you go over material lists and job schedules in your pre-job planning sessions, make a list of tools that you will need for each section's work, and when that is likely to be. By using this list, you can plan ahead and make sure the tools you need will be available when they are needed.
4. Make it a policy that whenever your men are finished with a tool on their jobs, they return it to the shop immediately. This will usually eliminate a lot of lost and stolen tool problems.

YOUR FOREMEN

You can make the best job management decisions in the world, but it's your foremen that must carry out your policies on the job site. For a contractor to properly train his foremen is one of the most important yet most ignored things he can do. Your foremen are there on the job sites every working day; they are the ones that give your company a good reputation or a bad reputation. They are the ones who either help or irritate the customers, architects, and inspectors.

Your foremen hold a lot of your company's future in their hands. Make the effort to make sure they run your company's projects the way you want them to. Remember though that giving workers this type of training is not the same as teaching them how to pull wire. They will have to learn how to deal with personalities, attitudes, and other intangibles. This is very different from bending pipe or pulling wire, and most foremen are never given even an introductory course in dealing with these things.

From the industry's beginnings, this kind of training has been neglected. What usually happens is that a competent electrician is given the job of foreman and then spends several years learning to do his new job by the trial-and-error method. Obviously this is going to cost the company money (both short-term and long-term). It is worth your time and effort to make sure that your foremen know how to be foremen, not just how to be electricians.

Always have a pre-job meeting with the foreman and estimator before the project starts. Spend as much time as necessary to thoroughly review the plans, specs, submittals, and schedule. A lot of answered questions before the job starts will avoid a lot of problems during the course of the job.

Table 3–2 shows a list of responsibilities for a foreman. Give your foremen copies of a list like this and review it with them to make sure they understand what you want them to do.

The foreman holds a very important position in your company, so if you have a good foreman treat him with respect, and count him as an

TABLE 3–2 List of Responsibilities—Foreman

1. Plan each day's work the afternoon before.
2. Instruct and supply your workers.
3. Supervise your men's work.
4. Carefully check all deliveries.
5. Inform office of all back orders.
6. Maintain a good attitude among the workers.
7. Meet with the GC's superintendent once per day.
8. Coordinate with the other trades.
9. Keep a daily diary.
10. Plan ahead.
11. Keep up on any documentation needed.
12. Maintain a friendly relationship with the inspector, architects, and engineers.
13. Notify the superintendent or office of any major problems with the job.
14. Coordinate with the office to schedule material deliveries and manpower requirements.

important part of the team. As much as possible, try to always give your foremen a decent year-end bonus, especially the superior foremen.

Documentation

It is very important for your foremen to have daily diaries and to be diligent about recording each day's events—even such details as weather can be very important. This provides legal documentation of what goes on at the job site. If a lawsuit ever does arise (and they get more common all the time) a diary is irreplaceable. Oral testimony can be just your word against theirs, while written documents are given a lot more weight in the courtroom. A sample diary form is shown in Figure 3–4.

Keeping up on documentation is one of your foremen's chief responsibilities. (In a lot of cases you may want to have your superintendent handle this, but it will be your foremen that have to alert the superintendent to a problem.) Any matter that is of real importance is worth taking the time to write a letter about. As prone to lawsuits as the construction industry is, it is worth your while to protect yourselves. Figures 3–5 through 3–7 show sample letters that protect the electrical contractor from job problems that are developing. Don't neglect writing letters like these; they can keep your company out of a lot of trouble.

Mr. George Wilkes
Harquist Construction Co., Inc.
7140 N. Kolmar Ave.
Salt Lake City, Utah 88041

Re: The Animal Science Bldg. at Northern Utah University

Dear George,

As you know, this job is running several weeks behind schedule. We want you to know our concerns about this.

Because we so frequently have to wait for other people to finish their work before we can do ours, we have sometimes been put in a very uncomfortable situation at the end of a project, even when it has been in no way our fault.

From the beginning of the job, up to this time, we have kept up on our work, almost without exception. To the best of our knowledge we have not been responsible for any delays at all.

Figure 3–5

As of today, we pulled two men off of your job, as there wasn't enough work to keep them busy. Right now, the painter and the cabinet installer are holding up our work, leaving us only enough work for the job foreman and two journeymen. Until these other trades catch up on their work, there is nothing else for us to do. If the others don't catch up in a week or so, we will have to pull another man off of the job.

What concerns us is this: The job is scheduled for completion in four weeks; the painter and cabinet man look like they won't be done for at least another week. That leaves us only two to three weeks to finish the job, which is nigh unto impossible for a job as large as this one. We figure that we will need at least three to four weeks to completely finish the project. We would like to be doing this work right now; but because of other people, we can't.

The purpose of this letter is to inform you of these facts, so that we don't bear the blame for delaying completion of the project. Because we are so often the last trade on the job, it often can seem that the delays are our fault; but, *we cannot be held responsible for delays caused by others whose work must be done in order for us to complete our work.*

Although I know that you understand this situation, I hope this letter will clarify the situation for you, and avoid any potential problems between our companies.

Sincerely,

James McGuire, Project Superintendent

Figure 3-5

Mr. Edward DeLacosta
E.J. Patterson & Assoc.
2141 Palm Blvd.
Sarasota, FL 33414

Dear Mr. DeLacosta,

I am writing regarding a problem that has developed on the Gilbert County Administration Building Project, for which your company is the Consulting Engineer.

Figure 3-6

During the course of the project, we have run into a conflict with the HVAC people, as to use of the mechanical chase on the west end of the building (between rooms #124 and 126, etc.).

It appears that the original design of this chase was adequate, but that a change that was made in the HVAC system has overloaded it. We have been told that a change was made in the size of some of the various ducts that are in this chase; the problem is that we don't think there will be enough room left for all of the conduits that we have planned to go in the chase. We have two 3″ feeders going from the first floor electrical room to the penthouse (for the elevators) that are supposed to go through the chase. Also, we have several other feeders that were to use that chase at different levels of the building.

Obviously the HVAC people have no choice but to use this chase for their ductwork, but this puts us in a different position. We are not sure if it will be possible for us to fit all of our feeders in the chase with the oversized ductwork. Even if we can fit them all in, we have no way to get in to any pull boxes that we may need. So we think that using the chase for all of our feeders (as originally planned) is out of the question.

We do feel that we have one solution to the problem. If we change the location of the feeders to the penthouse, we will have adequate room for our other feeders. We think that we could move these feeders to the stairway just east of the chase (between rooms #129 & 131, etc.). This would leave the two 3″ feeders exposed in the corner of the stairway, but we don't really feel that this would be any problem, as the conduits are to be GRC, and this stairway is not a critical area as far as cosmetics are concerned. Of course, there would be the additional expense of core drilling, fire seals, and a longer run; but right now we see no alternative.

Would you please come out to the job and review this situation; our job foreman, David Wolcott, will be glad to assist you. You can call David at 341-7069 (lunch time is your best bet) to make arrangements. If you could take a look at this problem, review it with the architect and get us an answer soon, it would be appreciated. The job is at the stage right now where if we don't know what to do in the next week or so, we may start causing delays.

Figure 3–6

Thank you for your help Mr. DeLacosta; if you have any questions, or if you would like us to prepare any cost estimates, please let us know, and we will be glad to accommodate you.

Sincerely,

Michael R. Horwaith, Estimator

Figure 3-6

Mr. John West
P.D. Robertson & Assoc., Architects
2101 N. Kedvale
Madison, Wisconsin, 55421

Re: Southside Public Library Addition.

Dear John,

We have been asked by Mike James of Wolcott Construction to put the lighting system for this project into service. He said that it would make the painting and trim work a lot easier to do and to check. Apparently you discussed this with him also.

This would, in itself, be no problem to us, as this work is basically complete, but we do have some concerns that need to be addressed before we can do this.

What concerns us is putting these systems into operation so long before the owner's final acceptance. As you know, by our contract, we must guarantee the job for one year from the owner's final acceptance. If we put these systems into service now, they will probably be in actual use for three or four months before the owner makes a final acceptance; this puts us in a more liable position than is specified by our contract. Of course, we expect no problems, but who can tell if a piece of equipment was manufactured incorrectly, etc.? What really concerns us is the possibility of replacing a transformer or panelboard after thirteen months of use, but still under warranty. Certainly, the odds of this happening are very slim, but we just don't want to wager on something like this.

Do you think that you could give us a letter, stating the acceptance of these systems as of the date they are put into

Figure 3–7

service? The things we are concerned about are light fixtures, etc., transformers, lighting panels, lighting contactors, and switching circuits.

We appreciate your help in this matter. Please understand that we are not trying to be "hard-nosed," but just don't want to take any risks.

Thank you,

Jim Cosworth, Superintendent

Figure 3–7

Theft

Theft can be a problem area for electrical contractors. It may include theft either by a company's own people or theft by others; both happen too frequently. One quick answer is to keep a good eye on all of your materials—to keep them locked up—especially the items that are more valuable or easier to sell. This includes large sizes of copper wire, circuit breakers, motor starters, and light fixtures. You won't always be able to keep everything locked up, but if it is possible, go to the extra effort. Another helpful thing is to have all power tools returned to the shop as soon as they are no longer needed on the job site. This prevents a lot of them from "walking off." Doing both of these things will eliminate most of the problem of theft by outsiders.

A far more intricate problem is that of theft by people within the company. Ideally, this could be avoided by only hiring people who are trustworthy, but in the real world this is very difficult to do, and even if people are honest when you hire them, there is no guarantee that they'll still be honest next year.

The most likely type of person to steal material is the electrician who is known to do side jobs. He is the likely candidate for grabbing four bundles of pipe and ten reels of wire when nobody is looking, and throwing them into the back of his van.

If you suspect such a problem, you are going to have to play detective a little bit. Watch your jobs without anyone knowing. Some contractors have resorted to hiring detectives and having them work as electricians to get a first-hand view of what was going on at the job sites.

While on the subject of theft, it is worthwhile to mention theft by bookkeepers. Because of this possibility, you should have an outside bookkeeper or accountant review all of your books from time to time. There are enough horror stories about theft by bookkeepers to prove that it happens, and also that it can run into a very large amount of money very quickly.

Standardization

As much as possible, try to have certain standard ways of doing things; for example, use certain colors of wire for certain voltages or switching operations. This can be especially helpful in residential wiring, where you can use standards of switch and outlet mounting heights, marking of wires, circuiting, and so on.

When you have a set standard that everyone follows, you eliminate a great amount of confusion when one person's work has to fit in with another's (which happens every day), or when one worker has to follow another in a certain area. It is no harder to learn one way than another, so make sure your workers all learn the same way. In the course of time it will save your company a lot of money (although it can be difficult to notice). Table 3–3 shows a standard of wire colors.

TABLE 3–3 Wire Color Codes

277/480v Systems		120/208 (240)v Systems	
A phase	Brown	A phase	Black
B phase	Orange	B phase	Red
C phase	Yellow	C phase	Blue
Neutral	Gray	Neutral	White
Ground	Green	Ground	Green
Switch legs	Red or Blue (twisted)	Switch legs	Brown or Orange (twisted)
Three-way travelers	Pink	Three-way travelers	Purple

Supervision

Your superintendent should visit your job sites as frequently as possible. Every day would be best, but certainly no less than twice per week, unless the project is out of town. The main things for your superintendent to look for on his job-site visits are these:

1. That everyone is working and happy.
2. To identify and solve any problems with the job.
3. To make sure the customer is happy with the work.
4. To make sure the work is being coordinated between your workers and the other trades.
5. To review the job in detail with the foreman at least once per week.
6. To handle the paper work for any change orders.
7. To handle any job correspondence.
8. To order any inspections that will be required.

Once a week the superintendent should discuss the project with the estimator who figured it. When the entire job is finished, the estimator, foreman, and superintendent should sit down and review the entire job in relation to the original estimate.

Small Jobs

Very small jobs need to be handled differently from large projects. For service jobs, a routine like this is usually the best:

1. Get a good description of the job from the customer. If necessary, have the superintendent stop by and look at the job before you send an electrician.
2. Come to a clear understanding with the customer of how payment is to be made. (C.O.D. is usually the best for small jobs.)
3. Make sure your service man has enough of the right kinds of materials and tools to complete the job.
4. Fill out a service-order form, and go over it with your service person to make sure he understands how the job will be handled. If the worker is to pick up a check before leaving, make sure that the worker knows this also.

For jobs that are a little larger than this, send the superintendent first to prepare an estimate. Then, after you have agreed with the customer on both cost and payment, follow steps 3 and 4 as you would for a service call. Make sure that you write POs for any materials, and charge them to the job (even if you write a PO to stock).

For any jobs where you will need several electricians, or for time and material jobs, make a separate job folder; but you needn't keep detailed records as you do for a large job.

Simplicity. Remember that simplicity is a key factor in any management structure. If your management system is too complex, not only will it be wasteful, but you will have a hard time getting people to do what you want them to. As long as you maintain control, the rule is: The simpler the better.

4

JOB ACCOUNTING

Job accounting (or job-cost accounting, as some call it) is critical to almost any electrical-contracting enterprise. Without it, there is no way to check the progress and outcome of individual jobs. Nor is there any way to identify problem areas of individual jobs. Most established electrical contractors have some type of job accounting system—it's a matter of survival. If a company doesn't know what is going on at its job sites, it leaves itself open to a host of potentially disastrous problems.

For many electrical contractors, a job accounting system is nothing more than a record of the costs of material and labor. While this is certainly better than no system, it lacks any good way of identifying problems. With just a little more effort than is required for a simple cost record, a sectional-type job-accounting system can be implemented which will be far more useful.

Sectional accounting refers to a method of keeping a separate accounting record for each section of a project. As pertaining to electrical contracting, the sections are designated as the various parts of the electrical installation, rather than different locations, as different floors of an office building.

There are also accounting systems that some electrical contractors use, which break down their accounting, not only by section, but by each type of material which is installed. As a generic term, these are called complex job accounting systems.

A sectional type of job accounting is almost always best all-around for an electrical contracting company, as it uses information that is fairly easy to gather and provides the contractor with almost as much information as he can feasibly use.

COMPLEX SYSTEMS

Some large electrical contractors use a very complex system of job accounting, but with few exceptions these systems are far too difficult and expensive to operate, especially for a small or medium-sized contractor. Further, the complex systems aren't any more accurate than a sectional type of system. They seem a lot more detailed, and therefore should be more accurate, but in actual use, they really aren't.

First, the complex systems are expensive to operate. There are extra clerical costs for gathering and tabulating all of the figures. While these costs are the first thing that come to mind, they aren't the ones that really hurt. The costs that really hurt are the ones that aren't so easy to see: the time when the electricians sit down and try to remember what kinds of material they put in that day, the time they spend trying to figure out how many feet of half-inch EMT they installed, or how many feet of two-inch GRC. Maybe they helped Mike and Rob pull a feeder run—it took them 2½ hours; they pulled in three #500s, one #2/0, and one #4bare; the whole run was 218 feet long—now, what do they write on the time card? This type of wasted time isn't obvious (they never fill in their time cards when the boss is around), but it is very real, and very costly.

Secondly, these complex systems are not much more accurate. They appear to be more accurate because of their great detail; but remember that the information you get out is no better than the information that you put in. These systems require a lot of input from the company's electricians (or sometimes the job foreman). The problem is that this doesn't often get done the way it should. Sure, it gets done right the first couple of times, or when the foreman raises a fuss over bad time cards, but what about at the end of a hard day in 90-degree heat? Or on a Friday afternoon when everyone wants to get as far away from the job as they can? It just doesn't work as well in real life as it does on paper. People get in a hurry and forget important facts. Even when intentions are good, it's just too much data for a crew of electricians to get correct.

Thirdly, any information you can obtain will need clarification before it can be used. For example, if there are 100 extra hours spent on two-inch EMT than were originally estimated for a particular project, are you going to say "Gee, our labor rates for two-inch EMT are only half of what they should be; lets double our labor rates."? No way. You're going to get a hold of the foreman and say "Hey, what happened with this two-inch EMT? It took twice as long to install as it is supposed to." Then the foreman will give an answer as to what happened. Then you will have some useable information. So, if the final answer will come through the foreman anyway, why take the time and expense to gather piles of information that will be overridden in the end?

It's not that the complex systems give you bad information, but that they cost a lot more and don't really provide information that is substantially superior.

SECTIONAL SYSTEMS

The sectional system is designed for simplicity. The office people do almost no more work than they would for a simple cost record; they just write the numbers down in different places. Your electricians only fill out a simple time card. All purchases and material invoices are checked automatically by the system.

For this type of system, the job is divided into several sections, which are accounted for separately. The best way to divide jobs into sections is as follows:

Section One	Slab, site work, and temporary power
Section Two	Interior raceway
Section Three	Distribution Equipment
Section Four	Wire
Section Five	Light fixtures
Section Six	Trim out
Section Seven	Special systems, or other

These sections are designed to correspond with the way electrical projects are actually put together, which makes the accounting more accurate and the information easier to gather.

Section One. *Slab, site,* and *temporary power* also includes any preliminary job set-up and the like. The term *slab* refers to the first-floor slab only, which is done toward the beginning of most projects. If a particular project you are working on happens to be a multistory building, all of your work in the upper floor slabs will be considered interior raceway (Section Two).

Section Two. *Interior raceway* is your entire raceway system. This is not limited to conduit only, but also includes the following:

Conduit fittings
Pipe nipples
Bushings
Straps
Service heads
Condulets
Outlet boxes
Pull boxes
Underfloor duct (where not included in Section One)
Wall duct
Wireways
Cable tray
Enclosures and cabinets
"Unistrut"
Miscellaneous

You will notice that miscellaneous is only included in this section. This is because Section One is usually the largest section of the project, and is the easiest and best section in which to include miscellaneous items. It is one of the first sections of the work that will be completed; by including them here you will get paid for these miscellaneous items toward the front of the job, rather than toward the end. Miscellaneous items include:

Plywood backboards
Pitchpockets
Tape
Welding
Ductseal
Solder
Paint
Name plates
Fasteners (screws, bolts, etc.)
Fixture clips
Lugs
Wire nuts
Ground rods
Grounding clamps
Grounding pigtails
Chain
Threaded rob
Strain relief grips
Saw blades
Any other odd items

Section Three. *Distribution equipment* is all of the switchgear that goes into the job, and similar items. This does not include very large items such as substations and overhead distribution; these things would be considered special systems, Section Seven. Distribution equipment typically includes the following items:

Switchboards
Panelboards
Loadcenters
Motor starters (magnetic)
Motor control centers
Circuit breakers
Fuses
Meters
Meter centers
Current-transformer cabinets

Transformers
Capacitors
Enclosed circuit breakers
Safety switches
Manual starters
Lighting contactors
Bus duct
Time clocks
Control items (photo cells, push buttons, etc.)

Section Four. *Wire* includes any and all types of wire and cable.

Section Five. *Light fixtures* include all light fixtures, lamps (bulbs), special hangers, track, poles, and so on.

Section Six. *Trim* includes all trim out and equipment connections. It also includes any final testing, maintenance instructions, and the like. The items that typically come under this classification are:

Switches (all types)
Receptacles (all types)
Cord sets
Trim plates
Plugs
Flexible conduit connections (Liquid tight conduit, liquid tight fittings, flexible conduit [Greenfield], flex fittings, etc.)
Low voltage relays

Section Seven. *Special systems* can include any special type of electrical or electronic system. Often these systems are installed by specialty contractors who bid their work directly to the electrical contractor. Some contractors do a lot of their own special systems work, although it is typically better to give the best of your efforts in doing your main job (installing pipe, wire, fixtures, and outlets) and letting someone else do the specialty work.

Certainly a good electrician can do specialty installations, but it may take him quite a while to get a feel for what he is doing. Think of it the same way as you would about sending an electrician to fix a washing machine. You send a good electrician; he opens up the machine, checks the schematic, and after an hour has it fixed. But a good appliance serviceman will listen to the machine operate and know in a moment what is wrong with it. He can have it fixed in ten minutes. The guy who only does the work occasionally is terrible compared to the guy who does it five days per week.

Special systems typically include the following:

Fire alarm systems
Nurse call systems
Sound systems
Central clock systems
MATV and CATV systems
Security systems
Special hospital systems
Emergency power systems
Energy management systems
Control and process systems
UPS systems
Lightning protection systems
Signaling systems (bells, buzzers, etc.)
Electric heating

Figure 4–1 shows a typical section accounting sheet. Notice that all entries are made weekly; this is designed to fit in with the schedules which were given in Chapter Two for the bookkeeper, estimator, and superintendent. The information that will be recorded on these sectional accounting sheets is the same information that almost all electrical contractors gather every week, but here it is arranged a little bit differently.

First of all, each of these sheets is for only one section of the job. This means that you will need seven of these sheets in a job folder; one for each of the seven sections of the project. At the top of the sheet are places for the name of the job, the job number, the section name, and the section number. At the bottom of the sheet are places for the original estimate of material dollars and labor hours, which are included for checking on the job's progress and for final review.

The first column (Week Ending) is used to identify the time period for the recorded figures. The second column (Hours) is for the number of labor hours used for that section's work that week. The third column (Hours to Date) is a total of all the hours used on that section's work from the beginning of the job through the current period.

Notice that labor hours are used rather than labor dollars. This is because labor is measured in hours rather than in dollars. Labor is estimated by assigning a certain amount of time for the installation of each piece of material on the job. Thus, labor is estimated in hours, not dollars. A dollar figure for labor expense is calculated by multiplying the total

Job Name________________ Job Number________________

Section__________ — ________________

Week Ending	Hours	Hours to Date	Material $	Material to Date	%	PO No.	Complete

Estimated Hours Estimated Material $ Final Review by

Figure 4–1

number of hours by a company's average hourly wage rate. However, if a particular company had a need to use labor dollars rather than labor hours, this sheet would still be used, with dollar figures used in the first three columns rather than hour figures.

The fourth column is Material Dollars. This figure is the total dollar amount of invoices received for that section of work for that week.

The fifth column is Material to Date. This is the total amount of material billed to this section from the beginning of the job to date.

The sixth column (%) is the percentage of completion. This column must be filled out by the superintendent or some other experienced individual, and should be based upon an actual job-site inspection. Note that this is the completion percentage for this section of the job, and not the completion percentage of the whole job. This is used for both for billing purposes and checking progress against the original estimates of material cost and labor hours.

The last two columns are used to keep track of purchase orders. To keep your operations under control, all of your purchases should be made with purchase orders. Even if the purchase order never leaves your office, you should still use the PO for control purposes. Without using purchase orders, things are eventually forgotten, misplaced, and incorrectly accounted for. This is the kind of chaotic office situation that you want desperately to avoid.

In the PO Number column you will fill in the number of any purchase order written for material for this section of the job. This will require you to write a separate purchase order for each section of the job; writing POs that include items for more than one section of the job will throw the accounting system out of order.

The last column is for checking off each PO when it is complete and correct. By doing this, you will be sure that all billing from your suppliers is correct, and that you pay exactly what you expected to pay.

Figure 4–2 shows a sectional accounting sheet that has been completely filled in. This sheet is for the interior raceway section of an office building project (Section Two). Notice how the material figures lag a week or two behind the labor-hour figures. This happens because it takes several days between the time the material is delivered to the job and the time that you receive the invoice and verify it.

As you look at the last week recorded (2/14), notice that $171 is deducted from the material cost (numbers in parentheses are deducted, rather than added). This is for material that was ordered and delivered to the job site but not used. It is returned to the warehouse, deducted from the material cost of this section, and returned to stock for use on another project. This is recorded as PO #1471-R, with R standing for return.

The reviewer's final notes on the section are toward the bottom of the sheet (the entire job is reviewed when it is complete). Material cost ended up being 2.7% higher than expected, and labor hours were 7.5% lower than expected. These figures are calculated by dividing the actual cost by the estimated cost. (Actual material cost is divided by estimated

Job Name Jones & Co. Office Job Number 8741

Section Two — Interior Raceway

Week Ending	Hours	Hours to Date	Material $	Material to Date	%	PO No.	Complete
7/4	157	157	—	—	10	0824	✓
7/11	164	321	19,820.00	19,820.00	20	0920	✓
7/18	144	465	—	—	35	0964	✓
7/25	141 ½	606 ½	424.61	20,244.61	55	1147	✓
8/1	150 ½	757	2,417.60	22,662.21	60	1471-R	✓
8/8	104	861	—	—	70		
8/15	92 ¼	953 ¼	1,171.15	23,833.36	84		
8/22	60 ¼	1013 ½	461.50	24,294.86	92		
8/29	20	1033 ½	141.84	24,436.70	95		
11/7	41	1074 ½	—	—	98		
11/14	27	1101 ½	859.16	25,295.86	100		
11/21	—	—	417.84	25,713.70	—		
2/14	—	—	(171.00)	25,542.70	—		

Material — + 2.7%
Labor — - 7.5%

Estimated Hours 1,190. Estimated Material $ 24,850.— Final Review by BP

Figure 4–2

material cost; actual hours are divided by estimated hours.) The reviewer signed the sheet and recorded the results of that section on a job review sheet.

For the sake of accuracy, be sure to include sales tax in all of your figures. This will be done automatically by your suppliers on all of your invoices, but you will have to add the sales tax into your estimated material cost yourself.

Figure 4–3 shows a job review sheet. This sheet is for a final review of the whole project, and includes not only the costs recorded in the sectional accounting sheets, but job costs also. Job costs typically include the following:

Storage trailer rental and insurance
Office trailer rental and insurance
Equipment rental
Job site utilities

Job Review Sheet

Job Name ______________________ Job Number ______________________

Reviewer ______________________ Date ______________________

	Hours	Material $	Est. Hours	Est. Mat. $	Hrs. Results	Mat. $ Results
Section 1						
Section 2						
Section 3						
Section 4						
Section 5						
Section 6						
Section 7						
Total						

Job Costs

Week Ending	Item	Amont	P.O. #	Complete

Estimated Job Costs-

Comments:

Figure 4–3

Blueprinting costs (As-built drawings, etc.)
Travel and lodging expenses
Performance bonds
Freight expenses
Any other expenses caused by the job, but not included in the company's overhead

Figure 4–4 shows a completed Job Review Sheet.

Job Review Sheet

Job Name GREENFIELD CONDOS Job Number 4602

Reviewer C. J. Date 4/24/87

	Hours	Material $	Est. Hours	Est. Mat. $	Hrs. Results	Mat. $ Results
Section 1	197½	1,217.84	210	1,100.—	−6%	+10.7%
Section 2	1,249	24,180.41	1,290	24,854.—	−3.2%	−2.7%
Section 3	452	12,580.—	420	12,580.—	+7.6%	=
Section 4	801½	8,754.32	785	9,182.—	+2.1%	−4.7%
Section 5	990	41,517.40	950	41,400.—	+4.2%	+0.3%
Section 6	701	4,090.—	480	4,280.—	+46%*	−4.4%
Section 7	—	6,400.—	—	6,400.—	—	=
Total	4,391.	98,739.97	4,135.	99,796.—	+6.2%	−1.1%

Job Costs

Week Ending	Item	Amont	P.O. #	Complete
11/12/86	TRAILER	250.00	3214	✓
11/12/86	BACKHOE RENTAL	1,474.—	2981	✓
12/12/86	PHONE	144.60	3291	✓
12/12/86	TRAILER	250.—	3214	✓
1/15/87	PHONE	121.40	3291	✓
1/15/87	TRAILER	250.00	3214	✓
2/18/87	PHONE	97.40	3291	✓
2/18/87	TRAILER	125.00	3214	✓
3/18/87	FREIGHT	17.84	3724	✓

Estimated Job Costs- $3,000.00 ACTUAL - $2,730.24

Comments: JOB RAN WELL & AHEAD OF SCHEDULE.

SECT. 6 (TRIM) HRS. ARE WAY TOO HIGH - REASON: ALL APPLIANCES (REFRIGERATORS, RANGES, WASHERS, DRYERS) CAME WITHOUT THEIR CORDS INSTALLED. SO WE HAD TO DO THIS WORK. (WE'VE SUBMITTED A BILL TO OWNER & ARE WAITING A RESPONSE.)

Figure 4–4

Figure 4–5 shows a time card which has been designed for a sectional accounting system. At the top is a place for the worker's name, the date, the job number, and the job name. Below that is a place for the workers to write their hours. Below this is a listing of the section numbers, just so no one ever charges their hours to the wrong section of the project. At the extreme bottom of the card are notes that identify the work week as Wednesday through Tuesday, and payday as Friday. This seems to be the best compromise between what the workers would like and what works out well for the company. By having the work week end on Tuesday, the bookkeeper is given two or three days to get everything ready so that the workers get paid on Friday, which most of them appreciate. Having Friday as payday is somewhat traditional; this tends to convey a stable, secure image of your company. Of course this can be changed; *but*, if you do change it, make sure that you also change the schedules that were given in Chapter Two. The bookkeeper's schedule, the estimator's schedule, and the superintendent's schedule will have to be changed.

A completed time card is shown in Figure 4–6.

DAILY TIME CARD

Name ____________________ Date ____________

Job Number ____________ Job Name ____________

Section	1	2	3	4	5	6	7
Hours							

Section 1—Slab, site, temp. Sect. 2—Interior Raceway
Sect. 3—Dist. Equipt. Sect. 4—Wire Sect. 5—Light Fixtures
Sect. 6—Trim Sect. 7—Other
Work week—Wed. thru Tues. Payday—Friday P.M.

Figure 4–5

DAILY TIME CARD

Name MIKE HARGER Date 2/7/87

Job Number 9610 Job Name DELTA HANGAR

Section	1	2	3	4	5	6	7
Hours	—	—	—	3	5	—	—

Section 1—Slab, site, temp. Sect. 2—Interior Raceway
Sect. 3—Dist. Equipt. Sect. 4—Wire Sect. 5—Light Fixtures
Sect. 6—Trim Sect. 7—Other
Work week—Wed. thru Tues. Payday—Friday P.M.

Figure 4–6

THE FLOW OF PAPER WORK

Controlling paperwork has long been a problem area for electrical contractors. To make this as easy and foolproof as possible, this sectional system is specifically designed to make the flow as smooth as possible. If you can process your paperwork (invoices, billings, payroll, purchase orders, etc.) smoothly, you can handle a tremendous amount of work with relative ease, but when the flow breaks down, it breeds hassles and confusion.

The key factors in processing paperwork smoothly are as follows:

1. *Having enough information.* Beyond just having the right numbers, this means being able to get the answers to important questions.
2. *Getting the information when you need it.* This is a particularly frustrating problem for bookkeepers; they get everything together to start processing bills, and then need one small question answered. So they call a job site and find out that the worker who has the answer is on the roof somewhere and can't be reached. So the bookkeeper has to put all of his materials away and start on something else. Then he will have to come back to his billing after he gets the question answered.
3. *Having enough people to handle the work.* Although this may sound elementary, it is very often overlooked. A lot of companies get overloaded and ignore the paperwork until people start crying loud enough. Then they drop everything else and do the paperwork as fast as they can. This creates bad feelings and inefficiency, and perpetuates a cycle of only doing what is unavoidable.

For these reasons, the schedules given in Chapter Two are designed to put the right people in the right places at the right times. You will notice that the bookkeeper is scheduled to do billings and collections on Fridays. This coordinates with the estimator and superintendent being together in the office at the same time. If the bookkeeper has any questions regarding the billings or collections regarding any of the jobs, he has the person with the answers right there.

The following is a list of exactly how the different documents should be routed, and how many copies there should be of each one:

Purchase orders. Purchase orders should be made in triplicate.

One copy will go to the vendor.

One copy will go to the job folder to be used to check all invoices. The superintendent and estimator are scheduled to do this on Fridays.

One copy will go into a PO file. This file should contain copies of all purchase orders written, in numerical order. This gives you a way to quickly check every PO, even if it is lost or misplaced.

Billings. In general, billings should be in triplicate.

One copy should go to the job folder.
Two copies should go to the customer.
The customer should return one copy with payment.

Time cards. Time cards to be in duplicate.

One copy should go to payroll.
One copy should go to the job folder, to be posted to the sectional accounting sheets.

Incoming invoices. You should get two copies of every invoice. Both copies together should be verified by the superintendent and estimator at their meeting on Friday. If they are correct:

One copy should go to the job folder, to be posted to the sectional accounting forms.
One copy should go to the bookkeeper's accounts-payable basket.

Incoming statements. Incoming statements should be in duplicate.

One copy should go to a "Paid" file, together with a copy of all invoices pertaining to it.
One copy should be returned with payment.

Purchase Orders

Figure 4–7 shows a sample purchase order. At the top of the purchase order will go your company name, not shown in this sample. At the top right will be the purchase order number, numbered by the printer that actually prints your purchase orders. Also to be filled in the top portion of the order are the vendor's name, date, and shipping instructions. The body of the purchase order is for a description of the purchase and its terms; refer to Figures 3–2 through 3–4 for examples.

At the bottom of the purchase order are three small boxes which will be filled in with a three-part purchase order number. In the first box will go the purchase number that is found in the upper-right corner of the form. The job number will go in the second box. The third box is for the section of the job for which the purchase order is written. Using this three-part number makes the job of checking invoices quite a bit easier and more foolproof. When the person that is checking invoices reads these numbers

Purchase Order

To ____________________ Date ________

Ship To ____________________

Shipping Instructions ____________________

Deliver The Following ———————— As Specified Herein

Submit Two Invoices
bearing this PO number

By ____________________

Received and acknowledged by __________ Date __________

Figure 4–7

on the invoice (make sure that your suppliers write the order number on all invoices; almost all suppliers willingly comply with this, but you may have to remind them), they can go right to the correct file and find the original purchase order without any delays. To the right of this is a place for the company name again, and a place for the purchaser to sign the order.

At the bottom of the order is a place for the vendor to sign, acknowledging that he or she received the purchase order and accepts its stated conditions. In today's lawsuit-minded marketplace you should have your supplier sign an acknowledgment for every purchase order that your company puts out. Even if the salesman seems to be a great guy, you don't know what his boss or his boss's boss are like. Problems like this occur all the time; do all you can to protect your company.

While discussing purchase orders, it is very important to mention that for the system to work correctly it is essential that your job foremen carefully check each and every material order that they receive. If your foremen don't go through every delivery ticket and verify their accuracy, there will be a missing link in the system. Make sure that your workers know how important it is to the company that they check each order carefully. Some people's nature is such that if you let them get away with something, they will tend to do it as frequently as they feel like doing it; so make sure that your people know that you are serious.

Electrical suppliers make as many mistakes as anyone else, so it is worth your while to check up on them. Remember that once a delivery ticket is signed by your people, your company is obligated to pay, as the supplier then has legal evidence that your company received everything that is written on the delivery ticket.

The Bookkeeper's Job

As we get into job accounting, the role of the bookkeeper becomes more important. The bookkeeper's schedule is given in Chapter Two, which shows which tasks will be done each day. On Monday, the bookkeeper will take the invoices that have been reviewed and approved by the superintendent and estimator during their meeting the preceding Friday. They will be recorded on the sectional accounting forms and put with the payables. On Tuesday, the bookkeeper prepares the accounts payable, and issues checks for the payment of bills that are due. Bills for the payment of items or services that are not part of specific jobs (office expenses, legal fees, publications, etc.) will need to be checked first. This can easily be done by referring to the purchase-order file, then verifying that the order matches the bill. The copy of the purchase order in the file should then be noted, dated, and initialed that it has been paid.

On Wednesday, the bookkeeper separates the two copies of the previous week's time cards into two piles. One copy of each time card will be used for computing payroll, with the other copy being set aside to be posted to the sectional accounting sheets later. The time cards should be checked by the superintendent first thing in the morning. Next, the bookkeeper will go through the standard duty of preparing the company's weekly payroll.

On Thursday, all of the week's payroll hours should be posted to the sectional accounting sheets for the various projects. Also, all of the invoices received that week should be separated by job. They should then be put in each job's folder so that the folders will be ready for the estimator and superintendent's weekly meeting. On Friday, the bookkeeper prepares billings for the company's projects. He should also take care of any past due bills that need to be collected.

The bookkeeper should have several files or baskets to keep paperwork in:

1. Accounts payable
2. Accounts receivable (billings)
3. Billings to go out (These are billings prepared by someone else, and given to the bookkeeper to process. The bookkeeper can give billings to the receptionist to type up and then check and record them as required.)

Methods of billing. There are several ways to handle billings, each with advantages and disadvantages. Every company is a unique organization, so there is no one method of billing that will suit every company. If your company does a lot of service-type work, it would probably be best to send out invoices for each service call, along with a separate statement combining them at the end of each month. On the other hand, if your company does almost all contract work, sending out only invoices would be a lot simpler, and therefore better.

The factors of taxes and overall accounting methods also come into the picture here. It is important that you spend some time talking to an accountant to decide what methods would suit your company best; but remember, the simpler the better.

SERVICE JOBS

Small service jobs can be very important to your company, but they need to be handled a little bit differently from large contract jobs.

Figure 4–8 shows a Service Order form, which is designed to account for all of the important parts of small jobs with a minimal amount of paperwork.

At the top of the form is a place for the company name; to the right of this is the order number. Just below the order number is a place for recording when the job has been billed. Whoever bills the job should initial and date the form on this line. Also in the top section of the form is a place to record the job information (customer's name, phone number, address, etc.).

The central section of the form is for the description of the job, including the electrician's information and the billing description. Of course, someone should discuss the job with the electrician beforehand, but it really helps to give a worker some notes to refer to. Just below this are places

Service Order

Billed____________

Customer- ____________

Address- ____________

Phone- ____________

Job Address- ____________

Job Phone- ____________

Description: ____________

Job Assigned To: Date: Completed:

Material		Labor		Other
Invoice P.O. #	Amount	Date	Hours	
	$			$

☐ T & M - Rate-

☐ Quoted Amount - $

Figure 4-8

to record to whom the job was assigned, the date it was given to the electrician, and the date that the job was completed.

The next section is for recording the material, labor, and any other costs for the job. Because of the way materials are handled and used on service jobs, there is a place included for a purchase order or invoice number. (Note the prefix S in the purchase order number shown in Figure 4–8. When this number appears on an incoming invoice, the bookkeeper will be alerted that this invoice pertains to a service job rather than to a larger contract job.) If you take material out of your stock for a service job, be sure to write a purchase order to stock, so you can keep track of the value of the material used on the service job.

Billing information is at the very bottom of the form. Boxes are provided to check off whether the job is T & M (time and material, also sometimes called cost-plus) or a lump-sum price (quoted). The labor and material rates for T & M work should be recorded, as should the quoted amount of any lump-sum projects. For example, for a T & M job, you may write down "35.00/hour, and cost plus 20%". This would mean that time is to be billed at $35.00 per hour, and that material is to be billed at cost, plus 20%.

This form should be printed in duplicate, with the carbon between the copies extending only to the bottom of the section containing "Job assigned to," "Date," and "Completed." By setting the form up this way, the second copy of the form can be given to the electrician and will contain no cost or billing information. The reason for doing this is not to be sneaky, but because the electrician really doesn't have any need for this information (unless he is to collect the bill before he leaves the job). There is a history of problems developing over matters like this. There are plenty of people who get upset over the company charging $35.00 per hour for their services when they only make $10.00 per hour. They don't think about the company's other costs; they only see 35 and 10. Giving people only the information that they need leaves no ground for problems like this to thrive in.

As you get the material ready for the job, be sure to note where the material will be coming from. If during the course of the job you need to get more material, be sure to note it on the Service Order form. Without these notes, it becomes very difficult for the bookkeeper to avoid mistakes. (The most typical mistake is to bill a job, and then to get an unaccounted-for invoice for materials later.)

After the form is filled out and the material is ready, give the second copy of the form to the electrician who will be doing the work. Then place the first copy of the order in a basket marked *Service Jobs In Progress*. When the electrician has completed the job, he should return the second copy of the form to the bookkeeper. The bookkeeper should then staple the first and second copies together, and put them in another basket marked *Service Jobs To Be Billed.* Material charges and labors should be posted to these sheets until complete; then the job should be billed promptly.

Once the job is complete, all of the labor charges should be posted to the sheets within about a week, but the material charges may take from one to two weeks. This is why it is so important that material orders are noted on the form.

As with other jobs, all of the cost reporting, billing, collections, and the like should be done in accordance with the schedules given in Chapter Two.

5

CUTTING COSTS

There probably isn't an electrical contractor in the world who wouldn't like to cut costs one way or another. The reason is simple: the fewer costs, the more money the contractor makes. There are three basic ways for an electrical contractor to cut costs:

1. Buy material more cheaply than estimated.
2. Increase labor productivity, and thereby pay less for labor than estimated.
3. Cut overhead expenses.

One of the biggest problems that contractors have had with cost-cutting is knowing where to stop. There is a fine line between a company operating at peak efficiency and a company that is working as hard as it can, yet still isn't getting everything done. Every operation of the company is important to the company's success. If any one operation doesn't get done, the whole company will suffer. So, as nice as cutting costs is, you need to take care not to cut your costs too much, or you will be worse off.

THE FLOW OF ELECTRICAL MATERIALS

The first way of cutting costs is buying material more cheaply. To get a good understanding of how to do this, it is necessary to first understand exactly how the entire electrical manufacturing and distribution industry works. Figure 5–1 shows how the flow of material goes from raw goods to the finished product in the contractor's hands.

First, the manufacturer buys raw materials from whatever sources he best can. These raw goods typically are copper, iron, aluminum, steel, chemicals (for making thermoplastic and bakelite), and rubber. After the manufacturer gets all of these materials together, they are assembled and fabricated until the product is complete. Then they are usually placed in a large warehouse for distribution, usually by truck.

From the manufacturer, the material is shipped to electrical distributors throughout the country according to each distributor's order. To keep contact with these hundreds of distributors scattered throughout the country (and sometimes the whole world) most manufacturers use a system of manufacturer's rep firms (rep is short for representative). Each representative works in a certain area promoting the manufacturer's products and handling orders. The rep companies (also called agencies) are almost always paid a percentage of the manufacturer's sales in their areas;

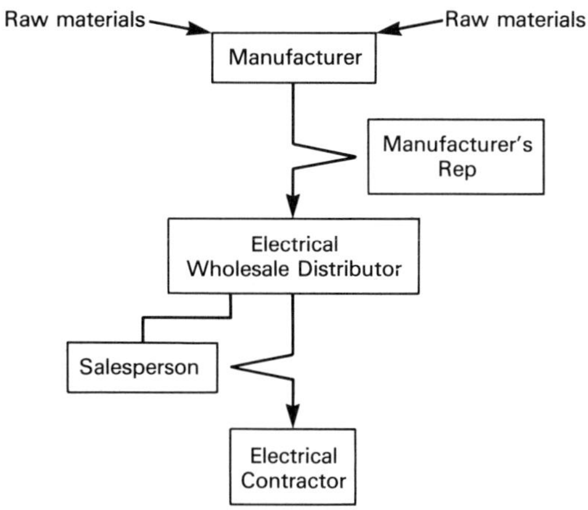

Figure 5-1

depending on the products, this is usually between four and ten percent of sales.

Manufacturer's reps typically fall into two basic categories, lighting reps and multi-product reps. Lighting reps specialize in lighting, and typically promote several different manufacturer's lines of lighting products. Lighting reps are expected to be proficient at doing specialized lighting layouts and calculations, so that they are not only salespeople, but also engineers to an extent.

Lighting reps spend a great deal of their time calling on architects and engineers. These people are the specifiers in the electrical construction industry; that is, they are the ones who specify what types of materials will be used on electrical-construction projects. Because of this, it is important for all reps (but especially lighting reps) to get to know the specifiers, and to make their products known to them. Ideally, reps like to get their products specified consistently by architects or engineers; because once a product is specified for the project, the installer will ultimately have to buy it. This is not to say that the over-the-counter trade is not important to lighting reps—it is—but the big sales (and big money) almost always come through the specifiers.

Multi-product reps typically promote other types of electrical products such as conduit, wire, wiring devices, fittings, fasteners, and specialty products. They rely much more on over-the-counter sales than do the lighting reps. Because of this, they spend a lot more of their time calling on supply houses than they do calling on specifiers. They want to see their products in stock at the local supply houses and being promoted to the local customers. These reps are interested in having their products specified by architects and engineers too, but they are not as dependent upon them as lighting reps are.

Once a rep gets an order from one of the distributors (the words distributor, supplier, and wholesaler are used here synonymously), he relays the order to the manufacturer, who usually ships and bills the order directly to the distributor. There are exceptions to this, but more often than not, this is how it works.

Once the material arrives at the wholesale house, it is put into stock in the wholesaler's warehouse to be sold. To sell their goods, wholesalers employ salespeople, who call on individual customers in their area. These customers are usually electrical contractors, industrial companies, institutions, large commercial accounts, or local electrical utilities, with electrical contractors typically the biggest customers. Electrical salespeople usually work on a percentage of profits. There are scores of angles to this, with the larger wholesalers using complex formulas to figure out the percentages. On the average, a salesman gets 12–14 percent of the profit on his sales, a company car, and some expense money, or he may get a flat percentage of profits (typically in the area of 20 percent), and pay for all of his own expenses.

Besides salespeople, a distributor also employs warehouse people, delivery drivers, counter people, managers, bookkeepers, and receptionists. The larger wholesalers also have quotations departments, and sometimes other specialized departments. On the average, an electrical wholesaler's overhead is between 14 and 17 percent on material that goes through the warehouse.

Wholesalers bill their customers monthly and expect payment within 30 days. Almost all suppliers will offer a discount to customers who pay by the tenth of the month. This discount is usually 2 percent, although on certain types of materials it can be 1 percent; sometimes there may be no discount at all.

There are some manufacturers who don't follow this pattern, most typically the manufacturers of distribution equipment, who employ their own sales force in place of manufacturer's reps. There are also a few manufacturers who sell directly to electrical contractors, and so make a higher percentage of profit, but they have a much harder time getting their products promoted and distributed.

Buying Material More Cheaply

The primary way to buy material more cheaply is to make it easier for your salespeople and supplier to sell to you. The salesperson doesn't want to spend hours calling on an electrical contractor and then sell material at a 10-percent markup. In this situation, the salesman might spent 10 or 12 hours, and only make about 20 dollars. To get the lowest possible prices, you need to make it as easy as possible for the supplier to sell you materials. Electrical contractors and electrical wholesalers are partners. In the long run, you have to work together. A supplier who gets pushed into a situation when he sells material to an electrical contractor at no profit is going to quote that contractor a higher price the next time around. As a contractor, put yourself in the supplier's position. He needs to make money to feed his family the same as you do, and he would like to make a reasonable profit just like you do. So the only long-term answer is to work together with the suppliers, rather than against them.

Making it easier for your suppliers. The first thing that will make it easier for your suppliers to sell you material is to not subject them to unfair competition. One of the electrical-supply business's biggest problems is bid-shopping by electrical contractors. In the typical scenario, the salesman gives what he feels is a very competitive price to a certain electrical contractor for a large order, after working several hours putting the bid together. The electrical contractor looks over the bid, and tells the salesman to check back in a day or two. The next day, another supplier's salesman walks in, looking for some business. The contractor lets him bid on the same materials as the other supplier. This second supplier calls back the next day with a price just slightly higher than the one the first supplier gave. The contractor tells the second supplier that the price is high, to which the second supplier replies "How about if I cut a thousand dollars off of the price?" Now the second supplier is low, and gets the order. Is this fair to the first supplier? Doing business this way has made some quick money for electrical contractors, while destroying their long-term business. Sure, the contractor gets a good price this time, but what about the next go-around? No supplier would give him their best price, knowing that their price will be used to help someone else get the order. From then on the quotes that this contractor gets are going to be high. On the other hand, if you treat your suppliers fairly, they will be glad to do you a favor every now and then. Electrical distributors don't mind fair competition, but if you engage in unfair competition, you make yourself their adversary rather than their partner. This business works both ways: The distributor needs your orders to survive, but someday you may need the distributor's help to keep yourself in business. If you were the distributor, would you bend over backward to help the contractor who is always shopping your prices?

Another way to make it easier for a distributor to sell you material at lower prices is to order the largest quantities you can. By doing this, the supplier does a lot less work per dollar of material sold, and can therefore sell that material with less of a markup. The more volume you do with the distributor, the lower the prices he can give you, and the larger the individual order, the lower the price.

One of the best ways to get lower prices is to get what are called *drop-shipments*. A drop-shipment is a direct shipment of material from the manufacturer to the contractor. When an order is handled this way, the supplier never touches the material; all he does is handle the paper work. The material never enters his warehouse, so his delivery people never have to load it and unload it from their trucks. Because of this, the markup that distributors charge on these orders is very small compared to the markup they must charge on material that goes through their warehouse. The drawback of drop-shipments is that they can only be done with large orders of material from one manufacturer. This is almost impossible for a smaller contractor, but even a medium-sized contractor can do this with light fixture and switchgear orders. The markup that suppliers charge for drop-shipped material can be as little as 4 percent; in some rare instances even lower. Compared to the 15 to 30 percent that is charged for material that goes through a supplier's warehouse, this is quite a discount indeed.

Another big help to suppliers is for a contractor to pay bills promptly. Suppliers go through a tremendous amount of material every month, which costs a great deal of money. If they don't pay their bills on time, the manufacturers will cut them off. This forces them to borrow from banks, which of course costs them a lot of money in interest charges. By paying them on time, you save the supplier from having to spend hard-earned profits on interest expenses. If at all possible, you should discount your bills. (That is, pay the supplier by the tenth of the month, deducting the discount that is typically shown on the statement.) This is a major factor to suppliers. A contractor who discounts his bill every month is like gold to an electrical distributor. (This is not to mention that the 2 percent discount you get by paying your bill 20 days sooner than you are expected to pays you a far better rate of interest than any bank does.)

Make things as easy as possible for the salespeople that call on your company. Look out for them. If a job comes your way that "has some fat in it," let the salesperson and supplier make a little extra on the job too; by doing things like this, you make the supplier want to do business with you.

How Much Should You Pay?

Table 5–1 shows a list of buyout percentages for various types of electrical material. As Note 2 says, these percentages are for medium-to-large orders, and are not for contractors who do a low-volume business. These are the percentages of the published contractor's price that you should expect to pay for these types of materials when ordered in a sizable quantity.

The prices are shown as the percentage of either the published prices from the Trade Service Corporation or from the Square D Digest. National Price Service's published prices are virtually identical to Trade Service's, and other switchgear manufacturer's prices are about the same as Square D's.

TABLE 5–1 Material Buyout Percentages

NOTES

[1]These figures show the percentages of the published contractor prices typically charged. All prices (except for distribution equipment) are referenced to the Trade Service Corporation's third column price (best price shown).

[2]These percentages are for medium-to-large quantities of materials ordered at one time. Suppliers should not be expected to provide these discounts on very small orders or to contractors who do a low-volume business.

[3]Distribution equipment prices are referenced to the Square D Company's annual Digest. (Shown with asterisk *)

Item	Buyout Percentage (%)
GRC conduit	60–67
GRC fittings	60–67
Intermediate conduit (IMC)	62–70
IMC fittings	62–70
EMT conduit	59–67
EMT fittings (steel)	40–50
EMT fittings (die-cast)	45–50
PVC conduit	42–55
PVC fittings	45–54
Flexible metallic conduit (Greenfield)	64–68
Flexible metallic conduit fittings	57–64
Seal-tite conduit	50–55
Seal-tite fittings	55–65
Pressed steel boxes (outlet boxes)	40–45
Plaster rings	52–65
Locknuts	40–50
Service heads	55–65
Pulling ells	50–65
Nipples	45–65
Straps	45–65
Bushings	37–54
Building wire	48–55
Cord (SO, SJ, etc.)	50–60
Type NMB Cable (Romex)	48–55
Fuses	72–80
Lighting fixtures	49–60
U Channel (Unistrut, Kindorf)	56–64
U Channel fittings	55–65
Ground Rods	65–70
Spec-grade wiring devices	65–75
Residential-grade wiring devices	52–57
Wire nuts	85–90
Time clocks	72–80
Residential loadcenters	30–35*
1 & 2-pole circuit breakers	30–35*
Other loadcenters	32–39*
3-pole circuit breakers	36–42*
General-duty safety switches (NEMA 1 & 3R)	35–41*

Item	Buyout Percentage (%)
Heavy-duty safety switches	42–48*
Under-floor duct	35–40*
Under-floor duct fittings	24–40*
Wall duct	72–77*
Wireway	40–47*
Wireway fittings	40–49*
Cable tray	40–45*
Cable Tray fittings	30–45*
Distribution-type circuit breakers	52–57*
Dry-type transformers	36–44*
Magnetic starters	45–54*

Trade Service Corporation
10996 Torreyana Rd.
San Diego, CA 92121

Square D Company
Executive Plaza
Palatine, IL 60067

National Price Service
4525 West 160th St.
Cleveland, OH 44135

Subscribing to one of the pricing services will keep you as up to date on pricing as your suppliers are, because the pricing services have access to the same information as the suppliers do, and they publish updates weekly. Almost every supplier in the country uses the pricing services for pricing much of their material. These published prices are shown in several columns, each column showing a price based upon a certain type of order (full box, over or under a certain dollar figure, etc.). In reality, any column except the lowest is useless except for the very smallest electrical contractors. Any contractor doing over $100,000 per year in total sales should almost never pay more than the best price shown in the pricing service books. Only contractors who do a very low volume and expect a lot of service should pay more than the lowest column price, except in special circumstances.

The lowest column in the pricing service books shows a price that includes at least a 30 percent markup for the supplier. The only exceptions to this are a few of the wiring accessory lines (cable ties, crimp connectors, shrink wrap) for which the published price only leaves 24–28 percent markup. At a 30 percent markup, the wholesaler's cost would be 77 percent of the price shown in the pricing books, to which the supplier must add a percentage of overhead and profit. At a discount of 10 percent off book price (best column), the supplier covers his material cost plus at least

17 percent for overhead and profit. (Actually, the supplier will get more than 17 percent most of the time, as the best column price shows *at least* a 30 percent markup, and often shows a price that reflects a higher markup.) For over-the-counter material purchases, this price level is usually a fair compromise between the supplier and the contractor. For certain items, many suppliers will be willing to go to less of a markup. This is particularly true for conduit and wire, for which the pricing services show prices that reflect a 65–100 percent markup. For large deliveries the supplier's percentage of markup should be less, as it requires less work for the supplier than having counter people wait on customers and go through the warehouse picking small orders.

Notice that all of these techniques for getting lower prices center around making it easier for the supplier to sell you the materials. Remember that the contractor and supplier are partners. When the partnership and trust between both parties is broken, the contractor and supplier start to work against each other rather than with each other. Ultimately this will cost both parties dearly. The electrical business is a small world; each part of the industry needs and is dependent upon the others; pulling together, we can survive well; pulling apart, we all get hurt.

COMMODITY ITEMS

There are certain electrical-construction materials that are more of a commodity than the others; that is, they are somewhat generic, and subject to greater price fluctuations than other types of materials. Commodity-type items include the following:

GRC conduit
IMC conduit
EMT conduit
PVC conduit
Copper building wire
Aluminum building wire
Types NMB and SEU cable

Notice that these items are very high-volume items, and that they are made from raw materials (steel, copper, plastics made from oils, and aluminum) that are subject to sudden price fluctuations. When the cost of these raw materials jumps or drops, the price of the electrical materials that are made from them will almost always take a sudden leap or fall.

GRC, IMC, and EMT conduit will obviously fluctuate in price along with the steel market, although the change is not always immediate, which is probably due to the manufacturers keeping large stocks of steel. Steel boxes also follow the raw-iron prices, although not as quickly or severely as conduit does. PVC conduit prices go up and down (often dramatically) according to the oil market. PVC conduit is one of the newest and most unstable of the commodity items, so if you are either working on or bidding a project that will require a large amount of PVC conduit, talk to a couple of suppliers and reps to find out where they think the PVC market is going before you commit yourself to a firm price. (Remember though, that some of these people will try to get you to *buy now*, even when it is not in your best interest.)

Copper wire and cable prices rise and fall along with the raw copper markets. If you watch the copper markets (any sizable newspaper's financial section or the Wall Street Journal will have this information), you will find that the price of copper wire will follow the price of raw copper by about thirty days. If the price of raw copper rises, you can expect the price of copper wire to go up within about thirty days. In the last several years, manufacturers and suppliers have been trying to pass the price increases on quickly, but there is still a lag of up to thirty days. (Notice that they pass along price increases quickly; they aren't always so quick to pass along price decreases!) There is one other characteristic of the copper-wire market that is important to remember: Wire prices almost always go up in early spring and down in late autumn. Obviously a wild copper market could change this, but this is the standard pattern. The reason for this is simple supply and demand. In the spring, construction work throughout the country tends to increase, which increases the demand for copper wire. In the winter, the amount of construction work throughout the country falls off noticeably, which greatly reduces the demand for copper wire, which brings the price down. The wire industry knows this, and readjusts their prices automatically.

Aluminum wire follows a pattern similar to copper wire, although it is not as predictable and subject to more frequent and larger fluctuations.

BID PRICING

The question of bid pricing is a real predicament to electrical contractors. (By bid pricing, we mean the material prices that an electrical contractor uses in the calculation of a bid price or quote.) One of the few ways that electrical contractors can make a little extra profit is by using safe material prices when they bid a job, and then buying the material cheaper than the material figure that the bid was based upon. This is one of the little mental games that contractors play to hide a little bit of money in their jobs. If a lot of contractors were to stop playing this game, they would most likely not add the little bit of extra profit to their bid tabulations, and the market would suffer. It would be "more correct" to price materials exactly where you think the prices will be when you buy them, and add a little extra profit to the bid tabulations, but it is far easier for most people to "hide" the extra money in the material figures.

So remember, if you use the lowest material prices in your bid tabulations, you have no room to move, and no extra slack in the job, which can really hurt if anything goes wrong.

LABOR PRODUCTIVITY

The second main area of cost cutting available to electrical contractors is in increasing labor productivity, and thereby cutting labor expenses. To increase labor productivity, there are two routes to take.

1. Use more efficient installation techniques.
2. Increase the workers' motivation, so that they are more productive and waste less time.

Both of these factors are important, and they both should be used. Since these factors are different one from another, they will be covered one at a time; first we'll look at more efficient installation techniques.

The Division of Tasks

The term *division of tasks* refers to the way installation tasks are allocated from one worker or group of workers to another. There are two basic ways that tasks are divided in the electrical construction industry:

1. By area
2. By type

Dividing tasks by *area* is most common. Dividing tasks by area means that the worker or workers are responsible for doing all of the work in a certain area. A typical example of this is to assign a certain crew to trim out a certain part of the job. In this instance, the crew will do all of the operations that are required to finish the trim work in that part of the job—pulling the wire, installing the wiring devices, installing light fixtures, and so on.

Dividing tasks by *type* means that each worker or group of workers is responsible for doing a certain type of work. For example, four workers will be assigned to pulling all of the wire in a certain section of the project; two other workers will be assigned to follow them and install all of the wiring devices; lastly, all six workers will work together to install all of the light fixtures.

There is very little question that dividing tasks by type is more efficient than dividing them by area. This is the concept that made Henry

Ford wealthy. While the other automobile manufacturers were building their cars one at a time, Ford built an assembly line. All of the engineering was already done for the whole process, all of the materials were in place ready to be installed, and each worker was assigned to a certain task which they did repetitively. As a result, there was no time wasted looking for parts and deciding what goes where. Each individual did only one task, at which they became very proficient. The bottom line is that within five years of producing their first Model T, Ford was putting out 200,000 cars per year. They could sell their cars for less money than the other manufacturers could, and made *a lot* of money.

The only problem with dividing tasks by type is that it takes a little more planning than dividing tasks by area. It is very easy for a job foreman to say "Hey Joe, you and Mike go trim out the north half of the building." It takes more effort for him to coordinate Joe's crew, Bob's crew, and Frank's crew together. Different types of work go faster or slower than others; thus dividing tasks by type requires more of the foreman's attention, but the bottom line is the same as it was for Henry Ford: You make a lot more money by doing it this way.

CONDUIT INSTALLATION TECHNIQUES

There are a number of techniques that can be used to speed up the installation of conduit. As with the division of tasks, it is easier for your job foremen to just tell the workers to hang the pipe, and let them do it as they see fit, rather than telling them a specific technique of doing it. And again, the bottom line is that you make more money by doing it the way that requires a little extra planning.

Any percentage of your conduit installation labor that you can save is very important, as the raceway installation (which is mostly conduit) accounts for the majority of the labor on your jobs.

One of the first steps toward speeding conduit installations is to have your workers' ladders set up as work stations, as shown in Figures 5–2 and 5–3. Figure 5–2 shows a typical six-foot step ladder set up as a work station. There are two small notches cut in the side rails of the ladder just above the second step. On the right side of the ladder are two hooks. One hook should be on the front side rail for a hacksaw. The other hook should be mounted on the rear side rail (the side that usually has rungs rather than steps) for hanging a conduit bender. Another hook should be mounted on the left side of the ladder for a canvas bag. In this bag will go various small parts that are needed for the installation; typically this would be fittings straps or fasteners. (For left-handed workers, this set up should be just the opposite.) By doing this, all of the necessary items for the installation are arranged to be at the electrician's fingertips, rather

Figure 5–2

Figure 5–3

than scattered around the job site. If fairly large hooks are used, the electrician can pick up the ladder with all of its appurtenances hanging from it, and move from one location to another quickly and easily.

Figure 5–3 shows how to use the notches in the side rails for cutting conduit. These notches should be cut in a triangular shape with one side flush with the top of the second step of the ladder. The conduit to be cut should then be placed into these notches. The electrician can then place his left knee on the conduit (assuming that the electrician is right-handed) and cut the conduit easily.

Slabbing

Slabbing simply means running conduits underneath the floor slab of a building. Running conduits in the floor slab rather than overhead saves quite a bit of money in both material and labor. The reasons are as follows:

1. Running the conduit in the slab uses up to 50 percent less conduit than it takes to run the same circuits overhead. Running from one receptacle to another requires only 2 two-foot vertical runs of conduit between outlets, for a total vertical length of 4 feet in each receptacle-to-receptacle run. By going overhead, a minimum vertical drop of 7 feet is required to each outlet, making the total vertical length for the same receptacle-to-receptacle run 14 feet, rather than 4 feet. Additionally, by running under the slab, you can usually get away with a more direct (angled, rather than straight and perpendicular) routing of the conduits.

2. Conduits run under the slab need almost no strapping. This eliminates a great deal of the expense for both material and labor for straps and fasteners. (Fasteners are screws, bolts, nails, anchors, and so on.)
3. PVC conduit is usually used for running under the slab, rather than EMT. This provides a savings in material cost, as PVC is cheaper than EMT. Additionally, PVC comes with integral couplings, which saves both material and labor expense.

The only problems with slabbing are that it requires you to install a separate ground wire, which you may not have had to do using metal conduit, and mistakes under the slab are often more expensive to fix than mistakes overhead.

Slabbing is somewhat of an art; to do it well takes a little more care and concentration than installing conduit overhead. Here are several tips on slabbing:

1. Double-check the measurements. This is particularly critical when trying to hit the middle of an interior wall. The outside walls are easy to hit, as they are well defined, but the inside walls are not often well marked before the concrete is poured, and sometimes the walls don't end up where they are supposed to be.
2. Always take your measurements from a permanent reference point, such as an outside wall, an expansion joint, or a floor joint.
3. Measure from both ways. When measuring to find an interior wall, measurements should be made from both ways. Let's say that the first measurement is made from an outside wall, and the location is marked. Next, another measurement should be taken from some type of permanent reference in the other direction. Frequently, you will find that the building isn't built exactly as it was supposed to be. At this point whoever is in charge of the construction of the building must be contacted to find out which way you should move the wall. Even though this in an extra step, it is far better than having to cut the slab to move your conduits.
4. You need to have workers on the job the day the concrete is poured. Concrete is very heavy and cumbersome to work with, so often the concrete finishers will knock something over in the process of spreading the concrete. Even if the finishers go out of their way to put whatever they pushed around back in place, their chances of getting it exactly correct are not too great. Have your workers wear boots like the finishers use and wade right in with them if necessary. Make sure that your installation is exactly as it is supposed to be. An hour later will be too late.

5. Make sure that all of your conduits are covered. Concrete is a lot easier to keep out of your pipes than it is to get out later. Pipes often get knocked down during a concrete pour, and some concrete will get splashed around during the pour. Duct tape is cheap and works well, but be sure that several layers are used so that the tape is not easily punctured.
6. Strap the conduits that go up an outside wall. If you can firmly strap these conduits and cover their open ends, you will almost never have any problems with them.
7. Make sure that the installation is double-checked. Have the construction superintendent check any locations that are questionable; you don't get many second chances with a concrete slab.

Pre-Bending and Standardization

This technique can save a great amount of time in conduit installations. When properly used, it can enable an electrician to install *many* hundreds of feet of conduit per day.

The first step is to mount all of the boxes in the area where the installation is to go. They should all be mounted at pre-set standard heights (a certain height for every receptacle, and a certain pre-set height for every switch, etc.). Next, all of the conduit drops from the ceiling to either a switch or receptacle can be bent at one time. The steps to follow are these:

1. Verify the mounting heights (to the top of the box) of the boxes, and the ceiling height and construction. You can then calculate exactly how long each type of conduit drop should be.
2. Place as many pieces of conduit as you will need so that one end of the conduits butts up against a wall. Make sure that all the pieces are even with each other. Now a line can be drawn across all of them at once, and they can all be bent to the exact same length, whatever length it is that will be correct for the installation.
3. Cut or notch any of the framing members that you have to. Do this all at one time, rather than one-by-one each time the electrician comes to a new location.
4. Install all of the various drops. Have the electricians wear some type of work apron, so that they will have all of the connectors, straps, and fasteners that they will need with them and won't have to spend time looking around for them.
5. Lastly, tie all of the conduits together overhead to complete the installation. If the job foreman (or the superintendent or estimator) goes through the plans and marks which drops will be handled this way, it will simplify the matter quite a bit.

Trim Carts

Making special carts for carrying all necessary trim materials (switches, receptacles, plates, wire nuts, grounding pig-tails, flexible conduit and fittings) can be a good time saver. The savings here (as with most of these labor-saving techniques) is that the workers don't have to spend time trying to find everything that they need—it is right in front of them. The idea is to mimic the assembly-line process as much as possible. Without question, electrical installations are vastly different from building automobiles, but the more often you can arrange your labor force to work in this manner the more money you will save.

One other thing that can be done to save you a lot of labor on trim out (if you can bring yourself to do it) is to get some little dollies. (A dolly is a piece of wood about one foot square with wheels on the bottom.) The electrician can sit on the dolly when wiring receptacles. By doing this, the worker doesn't have to get up, squat down, get up, and squat down; he can just scoot from outlet to outlet. One contractor who dared to try this said that he saved 40 percent of his trim-out labor by using the little scooters.

Working at the Best Time

Sometimes schedules can be difficult to arrange on a construction project, but try to do your work at the easiest possible time. For example, it is easier to pull wire before any drywall has been hung on the walls. When the electricians can see each run of conduit rather than just the openings of the outlet boxes, it makes finding the right conduits very easy.

There are many different applications to this principle in electrical construction. Find as many as you can. One very successful contractor is known to install all of his underslab conduit, install the boxes on the ends of the conduit, and even pull all of the wire before the concrete is ever poured.

MOTIVATION

Motivation is quite a complex subject. What motivates one person doesn't affect another at all. One person may need to be nurtured along, while another does better when pushed or challenged. So there are no pat answers to motivating people, but there are basic principles.

It is important to first understand what it is that makes your people work. The first reason is that they need money to support themselves and their families. They need to take care of their needs for food and shelter. But this isn't really a major factor to most electricians, who have very few problems with feeding their families. Getting beyond this, it is necessary to understand the factors that keep people in touch with their jobs—the factors that make the difference between a worker who is just putting in time, and a worker who takes pleasure in being good at the job.

To start with, refer to Table 5–2; it shows the results of a study that was done several years ago. The first part of the study shows what a group of employees said they wanted from their jobs; the second part shows what their supervisors *thought the employees wanted.*

Two things in particular stand out about this study. (1) Supervisors don't really understand their employees, and (2) More important, people want their jobs to make them feel good. The time is past when a stern warning and a kick in the seat of the pants were all the motivation that a worker needed. Times change, and so have the things that motivate workers.

The key to getting the most from your workers is to instill in them the will to perform; that is, to make them want to perform not just because you demand that they do, but because they *want to.* Try to set goals for your people to reach. Be careful not to make them too easy or too hard. (If they are too easy, they don't mean anything, and if they are too hard, no one will bother to try.) People like to make accomplishments; give them something to work for. This may be as simple as having a certain section of a building completely roughed-in by a certain date. What it is doesn't

TABLE 5–2 What Employees Said They Wanted from Their Jobs

1. Full appreciation of their work
2. Feeling in on things
3. Sympathetic help with personal problems
4. Job security
5. Good wages
6. Interesting work
7. Growth in the company
8. Loyalty to workers
9. Good working conditions
10. Tactful disciplining
What Their Supervisor Thought the Employees Wanted
1. Good Wages
2. Job security
3. Growth in the company
4. Good working conditions
5. Interesting work
6. Loyalty to workers
7. Tactful disciplining
8. Full appreciation of their work
9. Sympathetic help with personal problems
10. Feeling in on things

really matter; what is important is to inject a little bit of fun and excitement into your electricians' work lives.

It is also important that you keep your people working together as a team. For any firm larger than a one-man company working together with other people is not just beneficial, it is critical.

Recognizing Family Ties

It is important to remember that most of your workers have families. Your people depend heavily upon their families, and their families depend heavily upon them. Because of this, you need to consider not just your individual workers, but their families also. You need to keep not only your worker happy to be working for your company (we will use a male worker as an example, as the number of male electricians vastly outnumber female electricians), you need to keep the worker's wife happy that he is working for your company. Don't underestimate the influence that a man's family (especially his wife) can have upon him; the man lives with these people seven days a week, 365 days a year for who knows how many years.

Like your worker, the first thing they will want out of his job is for it to meet their basic needs. After that, they are likely to be most concerned about how the job affects their family situation.

Figure 5–4 shows an example of one of the best techniques ever used to get a man's wife to appreciate his working for your company. It is a personal letter from the top man in the company to the worker's wife. The letter is to let the wife know that the company appreciates her support for her husband. The actual content of the letter can be changed according to the situation, but the overall feel of the letter should stay more or

Dear Jill,

I want to express my personal thanks to you for your support of your husband Frank for these last few weeks. As you know, we've been working overtime to finish the Loading Docks at South Jersey Harbor. I know this has caused Frank to get home late from work, which must have taken away from your family time.

Because of this, we are sending a small gift certificate to the wives of all the men who have been working on this project. Buy yourself something nice; you deserve it.

And again, thank you for supporting Frank during this very busy time.

Arthur A. Selmer, President

Figure 5–4

less the same. Always try to include a small gift certificate with the letter. Get gift certificates from a very nice store in your area, so that the wife will get something nice for herself. A gift certificate in the amount of one or two hours of her husband's pay will pay you back many times over.

For unmarried men, you could send a letter to their girlfriends (and change the wording slightly), or just send them their own little thank-you note.

This type of letter turns the wife's perception of the boss from that of an ogre getting rich off her husband's sweat to that of a kind, understanding person. It works.

Family picnics are another way of giving a man's family a little extra pleasure out of his job. Avoid wild Christmas parties and events that involve drinking; they cause far more problems than they are worth.

The A-Team Idea

One of the more interesting motivational ideas to come along is the A-Team concept. (A-Team meaning the best ones, the all-stars.) The idea of the A Team is to choose a select crew, and assign them to a certain fairly difficult project. On the appointed day, the A Team will show up at the specified job, and try to knock out all of the specified work by the end of the day. Here is an example: On Monday the superintendent places a notice on the company note board. The note reads as follows:

> This Friday, the A Team will be working at the Western University Gymnasium project. We are to rough-in the entire locker room area. After work we will be going to Georgio's restaurant for dinner. The A-team roster will be placed on this board Wednesday afternoon.

Doing this injects some fun and excitement into your electricians' jobs. Being part of the A-Team gives them a special challenge. Also, after they work hard on Friday, they go to dinner at the company's expense, which is always nice. The projects that are chosen for the A Team should be difficult, but attainable. The superintendent should always pick up his tools and work right with the men. If the estimator or owner could come too that would be all the better. Let the bosses get right in there and sweat as hard as any of them. It will really help your people's attitudes, as well as get a lot of work done in a short period of time. The A Team should always work very hard, with no breaks (except lunch) until all of the work is done.

Make sure that everyone in your company has plenty of opportunity to be on the A Team; leaving people out makes them feel badly. Change the roster every week and give everyone a chance. If you have workers who are trying especially hard, make sure that you show them that you notice by including them in the A Team.

OVERHEAD AND MANAGEMENT EXPENSES

Cutting overhead and management expenses is something that must be done with care. Certainly it is nice to cut expenses, but it isn't nice to cut necessities. Historically, electrical contractors have tended to cut their overhead and management expenses too drastically, which results in things that are supposed to get done not getting done, which hurts a lot.

The only overhead expenses that should be cut are those that don't contribute to the company getting everything done efficiently.

Computer systems are a tool that a lot of contractors have gone to for cutting overhead expenses. There is no question that a computer can do many office functions faster and cheaper than people can. However, care must be taken to get a computer system that fits the company's needs. The computer is supposed to serve the company; the company doesn't serve the computer. Also, anyone thinking about investing in a computer should be aware that there are a lot more expenses in buying a computer system than just the purchase price. Setting up a computer is a very intense task; you will need to allow plenty of extra time and money for it.

The management and accounting ideas in Chapters Two, Three, and Four will probably do more for trimming your company's overhead and management expenses than anything else (even a computer). The important thing is that your operations are streamlined and efficient, while everything that needs to be done actually gets done.

SUMMARY

Every company is different and has its own characteristics, be they good or bad. Where costs can be cut in any particular company is a question that the company's management must carefully consider. You can use the ideas presented in this book as a guide, but remember that everything that you decide to do must be done through your company's people. Consider their particular characteristics when making decisions.

6

SERVICE WORK

With the proper circumstances, service work can be some of the most profitable work that an electrical contractor can ever do. By the term *service work* we refer to trouble calls, electrical repairs, emergency service, and so on. However, the service business is very different from the typical electrical-construction business, and requires a different type of set up. A typical electrical contractor who does a lot of new construction usually does better to set up a separate service branch rather than trying to integrate a high volume of service into his construction based organization. The main attributes of service contracting are these:

1. **A larger base of customers.** The more customers you have, the more potential you have. It is notable that all of these customers have friends, and friends of friends. This produces a lot of referral work, which is nice work to get. Small jobs lead to big jobs; the more customers, the better.
2. **Excellent cash flow.** For most residential service calls, the money will be collected before the electrician leaves the job. This means that the money is in your bank before you ever get a bill from a supplier or have to write a payroll check. For most commercial accounts, billing is made monthly, and payment is usually made promptly, so the money is in your bank before you have to pay your supply bill.
3. **Extra profits on labor.** Almost any service contractor has a minimum service charge. In smaller cities the minimum is often 1 hour; while in large cities, the minimum is often 1½ hours. This means that there will be a minimum of 1 or 1½ hours of labor charged to the customer, regardless of how long the actual repairs take. This charge is to compensate for driving time, etc. from job to job (which takes longer in a large city, hence the greater minimum charge). If the contractor can arrange the calls efficiently, several hours of extra labor charges can be made on a good day. (As with the other ways of making extra profit, this requires a little extra coordination, but pays off handsomely.) For example, if the electrician can run 7 service calls on a good day (this is not uncommon), the total time billed to customers will be 10½ hours (7 x 1½). The electrician gets paid for 8 hours, and the contractor's overhead charge is also covered in the first 8 hours of billing, so everything beyond 8 hours of billing is pure profit. Assuming an hourly rate of $30.00, this is $75.00 per worker per day every time this can be done. This can add up to a lot of money in a hurry.
4. **High profits on material.** Typically, service work allows for a much higher profit on materials than other types of electrical work. For

a service contractor to double the cost of material is not at all uncommon. Routine maintenance accounts and commercial accounts are usually charged somewhat less than this, because of the larger amounts of work, but the percentages of markup are still far greater than they are for new construction.

REQUIREMENTS

Service work is not for every contractor, nor can it be done well in every location. It is important to make a bold distinction between service work and new construction work. Certainly they both involve installing electrical materials, but the similarities don't go much farther than that. There are a number of requirements that must be fulfilled before your company can get deeply into the service end of the business:

1. You must have a large enough market to sustain a service business. Usually this requires a population of at least 250,000 in your area. Seldom can an extensive business be carried out in an area with less of a population than this. An electrical contractor in a town smaller than this can do quite well on service work, but won't be able to do *only* service work (or even the majority of his business). Even in the areas that have a population of 250,000–500,000, if there is much competition it can be difficult to get steady work. In the largest cities however (New York, Chicago, Los Angeles, Boston, Dallas, Atlanta) a service contractor can do very well (some have become very wealthy).
2. You need to have electricians that can handle the pressure that service work places upon them. This is a very important consideration, as most electricians (especially those who are used to doing new construction) cannot handle service work for more than a day or two. A service electrician has to be experienced, capable of thinking on his feet, and somewhat of a diplomat. He will often be expected to walk into a building that he knows nothing about, find out what the problem is (often from an upset customer who knows nothing about electrical work), and then fix the problem immediately. (Remember that this customer is paying what he thinks to be an outrageous hourly rate, and follows the electrician around to make sure he works quickly.) Then, even if the electrician can get this done satisfactorily, he has to collect the bill before he leaves. A good service electrician can be hard to find.
3. You need to be prepared to handle more paper work than other electrical contractors. Service work requires more billings, pricing, and collections than new construction. The actual mechanics of how to

do this are explained in Chapters Two, Three, and Four; but you must be prepared to do a lot of it, which often requires an additional employee in the office.

4. You will need a good estimator/salesperson to handle the large number of estimates you must make and quotes you must give out. Again, this person must be experienced, able to think on his feet, and an expert at public relations. He must be able to diagnose the customer's problems, find a solution to the problems (hopefully one that the customer can afford), and convince the customer to let his company do the work. As with a good service electrician, this person is not always easy to find.

Getting Work

For most types of service work (especially residential) the best type of advertising is still the telephone book Yellow Pages (provided it is done well). For a service contractor, the design of a Yellow Pages ad is one of the most important factors affecting his sales. The basic idea is to design an ad that stands out from the rest. You want an ad that captures people's attention, but not by gimmicky means. Your ad should get people's attention, and at the same time purvey a professional image. When you are doing the design of your ad (take plenty of time, and do it right), run down to your local library, and look through several out-of-town phone books. See what kind of ads service contractors in other locations are using; you can get a lot of good ideas this way. Also remember that in advertising, creativity is very important.

Aside from the design of your ad, you will have to decide how large you want your ad to be. Yellow Pages advertising can be quite expensive, but the larger the ad, the better draw it will have. Content and creativity are more important than size, but you need the ad to be large enough to be seen. It is also very hard to be creative with a two-inch by two-inch ad.

Figure 6–1 shows another very effective form of advertising for a service contractor, a sticker that is placed on the panel cover(s) wherever the company works. It can be surprising how much work these stickers can bring in. If the sticker is made out of good material it will last for years, even decades. If there is ever a problem with the wiring at these locations, the people may go to the panel to see if a breaker is tripped and then they will see your sticker. The odds are that they will call you before they will call someone else.

As with Yellow Pages advertising, spend plenty of time designing these stickers. Don't use the standard type of stickers that printers and office suppliers sell; have your own custom-made stickers. Spend a little extra time and money; it will be far more effective. Be creative, using more than just one color. Remember, these stickers are one of your biggest

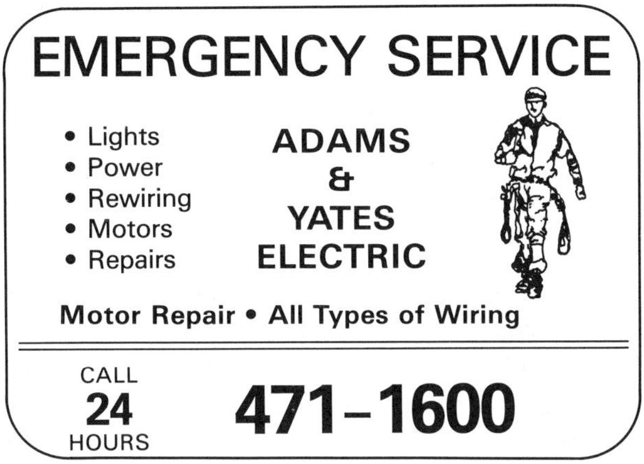

Figure 6-1

sources of contact with the public. Make the best impression possible with them.

For getting commercial business, having an outside salesperson usually works very well. The salesperson should call on all types of commercial and industrial accounts, explaining your company's services and offering free estimates. Because of the duties of this job, this person must be a proficient estimator as well as a salesperson. The percentages of commercial accounts that will buy after a salesperson calls on them are surprisingly high. When salespeople stop into businesses, they frequently get a response something like this: "Oh, I'm glad you stopped in; I've got a couple of things that really should be fixed. I've kind of been putting them off for a while. Take a look at them and tell me how much it will cost to get them fixed."

One big plus in selling to commercial and industrial accounts is offering a motor-repair service. Even if you just repair the smaller motors in your shop, and sub out the larger motors, it is a major selling point to many customers.

What you are selling your customers is service. This means that you take care of *any* problems, whether they are within the normal scope of electrical contracting or not. If the customer needs a new cord on their vacuum cleaner, put it on. If they need a lamp repaired, do it. These people don't care if it is electrical construction work—they are paying you to fix their problems, whatever they may be.

The phone-solicitation idea. The following is a method that a large service contractor in one of the nation's largest cities used to get a lot of business.

The contractor took a section of his office and had it remodeled, setting it up as a phone room. The walls were soundproofed, and a number of phone lines, desks, and chairs were installed. The contractor then hired several people to be phone solicitors. They went through the phone book calling people to see if they had any electrical work that needed to be done. Whenever they had someone who had a need, they would send one of the company's salespeople to give a free estimate. They got so many leads that it was difficult to find enough qualified salespeople to call on all of the interested customers.

Beyond just making contact, this company had one very important selling point: They would finance the work. They had made arrangements with several banks and finance companies so that they could get financing approved in just one or two days. Think about the benefit that this gave to the average customer. The average working person would probably have a pretty hard time shelling out enough cash for a new 200-amp service and a couple of outlets. By buying from this contractor, he could spread the payments over a year, and get the work done before bad wiring threatened to burn his house down. The contractor charged plenty for his services, but he made it easy for the customer to buy from him. He also did top-quality work. As a result, this company's biggest problem was trying to keep up with all of their work.

Doing the Work

Doing service work will force you to have more work trucks (or vans, which are usually better because they keep all of your tools and materials locked up and out of the rain), one or two more office people, and more material in stock. You will also need at least one person (or more) in your company who is an expert at almost everything electrical or mechanical. You will be running into a wide array of situations and, as the problem solver, you will be expected to have the answers. It is more than likely that you will need to have a full-time delivery driver to bring unexpectedly needed materials to the electricians. The material requisition form shown in Figure 6–2 is especially well suited to service contractors. The extra pricing columns on the form are handy for writing down both the book price and the selling price of the materials.

There is a question of what size crew to send to a service job—one person or two. Obviously this depends on the type of call, but from a profit-and-sales standpoint, it is usually better to send only one. This way the customer only has to pay for one worker rather than two, which can be expensive. Also, the contractor can get a little bit more for the hourly rate, and won't lose as much money if the customer isn't at home or decides not to have the work done (which does happen sometimes).

Doing service work will require you to work with a very large number of people. Be forewarned that you will run into all kinds of people, some

Material Request

Job Name: ______________________________

Address: ______________________________

____

Job Number: ______________________________ Job Phone: ______________________ Date: ______________________________

Ordered by	When Needed	Equipment Needed (Hand truck, Dolly, etc.)	P.O. Number	Filled by	Shipped Via	

Ordered	BO	Description	Shipped						
Received by:		Date:							

Figure 6–2

good, some bad. Don't expect to collect every bill that you send out; barring a miracle, it won't happen. Sometimes you might have to be kind of tough with a certain customer or electrician. This is all part of the service game. But if you are willing to play it, and learn how to play it well, there is no other type of electrical contracting that can pay you as high a percentage of profits.

7

MAINTENANCE AND SPECIALTY WORK

Maintenance (commercial, institutional, and industrial) and specialty work, like service work, offers a larger percentage of profits to electrical contractors than does new construction. But also like service work, maintenance and specialty work are very different from the typical electrical-construction work that most electricians are familiar with.

MAINTENANCE

Electrical maintenance is a multi-billion dollar per year market that offers a healthy percentage of profits, but there are a number of difficulties with doing maintenance work on a large scale.

1. Almost all maintenance work is done on an as-needed basis. This means that there is a tremendous amount of either feast or famine inherent to the business. This can make scheduling of manpower almost impossible for a non-union contractor, and very difficult for a union contractor.
2. It requires a large investment in special tools and equipment. A contractor who does a lot of industrial maintenance work can very easily have over $100,000 worth of special tools and equipment. This is obviously a big investment for tools that will spend a lot of time sitting around the shop.
3. It requires better-trained electricians. Knowing how to bend conduit and pull wire won't do it for maintenance work. A good maintenance electrician must have a thorough working knowledge of motors, controls, power distribution, and high-voltage work. Also, in the last ten years the need for the maintenance electrician to know electronics has risen. A maintenance electrician must be able to work and deal with people while under pressure; often a contractor has to train his own maintenance electricians, as they are frequently not available.

On the positive side, maintenance work keeps the contractor in close contact with the customer, giving the contractor the inside track on any new projects that the customer may have. Maintenance work allows for a larger percentage of markup on materials than does construction work. Contractors can usually buy electrical materials for 10–20 percent over the supplier's cost, while industrial accounts typically pay 30 percent above the wholesaler's cost. The maintenance customers are used to a higher price on materials.

If maintenance accounts can be obtained where the contractor has one or two (or hopefully more) workers at the customer's facility every day, the situation can be almost ideal. The feast and famine problem is removed, and even at a low hourly rate, this type of arrangement is great for the contractor, as it provides a stable cash flow and work force.

One big advantage to maintenance work is that most of it is sold as T & M (time and material) work. The advantage to T & M work is that it is very safe for the contractor. Aside from the risk of not getting paid, most of the inherent risks of electrical contracting are eliminated. Even if the percentage of profits seems lower (although it usually doesn't) the safety is invaluable.

Maintenance by Contract

The idea of contract maintenance has a lot of potential benefits, but has not caught on to any large extent. The idea of maintenance by contract is the same as that of a service agreement that you may buy when you purchase a major appliance. In return for a certain monthly payment, regular service is performed on the piece of equipment. Certain types of repairs are done for no charge, and others are done at a specified hourly rate.

Plant electrical maintenance does not differ from this, except that it covers many more pieces of equipment. The reason that this type of service has not caught on very well is that any large industrial company can afford to hire its own electricians. By hiring the electricians and giving them some training in industrial maintenance, the same men can do the same job, for just the cost of wages and a minimal start-up investment. This is also the reason that two-thirds of all industrial maintenance is done by in-house personnel rather than by electrical contractors. Maintenance by contract is a very stable and secure type of work for electrical contractors, but so far has not been very plentiful.

It is worthwhile to note that the reason that one electrical contractor is chosen over another for maintenance work is almost always *quality*. When these people need an electrician, price is not the important point; quality and service are what matter, and price runs a very distant second.

Figure 7–1 shows a sales letter for an electrical contractor who specializes in industrial maintenance and retrofit work. In selling industrial maintenance services (commercial and institutional also) establishing credibility is essential. Maintenance customers are extremely concerned with quality. These customers have plenty of need for electrical work; their real decision is who will *do* the work, rather than *if* the work will be done.

To run a maintenance business properly, the emphasis must be particularly strong on the long term. Customers must be handled carefully, so that they feel like they are getting an honest service at a reasonable cost at all times. Repeat business and steady customers are what makes money in the maintenance business, so this must be given top priority.

ABC Electric, Inc.
624 Broad Street
New Castle, DE 09144
(604) 274-6190

Dear Customer,

We would like to introduce ourselves to you. We are a customer-oriented electrical-contracting company, and we would like to service your electrical needs. We are also enclosing our brochure which describes the services which we have been providing to customers like you for the past 12 years.

Recently we decided that we would like to expand our customer base by 10–15%, so we have chosen several of the more respected firms in town that we would like to introduce ourselves to. We would like to start by doing a free walk-through of your facility. The purpose of this walk-through is for us to get to know your facility and its particular needs. If we see any needs in your facility, we will bring them to your attention with no strings attached. We don't mind investing some time and effort in potential customers, nor do we expect to make a large profit in every job that we take. We make our money over the long haul, by earning our customers' repeat business.

Our specialty is industrial and large commercial maintenance and upgrading; we do no residential wiring and very little new construction. Our company is specifically geared for plant maintenance, lighting retrofits, energy management, process control, and the like, which we feel we do far better than the average electrical contractor.

If you would give Jeff Rodgers (our general superintendent) a call at 274-6190 we will set up a time for our walk-through of your facility; the whole walk through should only take about one half-hour of your time, and, we would like to repeat, no strings attached.

We look forward to hearing from you.

Sincerely,

Johnathan C. Roche, Pres.

JCR/ns
enclosure

Figure 7–1

SPECIALTY WORK

Specialty work is almost always more profitable than regular electrical installations on a percentage basis. The real question is if there is enough specialty work available. This is an important consideration, since getting into a specialty market usually requires a sizable investment of both time and money.

There are two divisions of specialty work:

1. Special types of installations
2. Special service to specific types of customers

The first division, special types of installations, usually refers to specialized electrical, electronic, or mechanical installations. Any of these projects may be profitable, but it depends upon the local market whether there will be enough work of a given type in the contractor's local area to justify the expense of getting set up. These types of installations typically include the following:

Telephone work
Fire-alarm systems
Security systems
Television systems (closed-circuit TV, etc.)
Lightning-protection systems
Special electronic systems (nurse call, etc.)
Traffic lighting installations and maintenance
High-voltage work
Lighting-maintenance services
Energy management and programmable controllers

The second type of specialty work is that of providing a special service to certain customers. There are a number of industries that have very special needs for electrical installations. Industries such as the motion-picture industry, the printing industry, and others often need special types of wiring that most electrical contractors are not familiar with. Because of this, these customers look for the few electrical contractors who are familiar with their kinds of needs. In an area where this kind of work is available, it can be quite profitable. The hard part is trying to break into the business. Credibility is very important to these people, so trying to establish yourself as capable doesn't always come easy; it requires time and effort. The real question here is: *Is there enough of a specific type of work in your particular area to support this type of specialty business?*

WORKING WITH MANUFACTURERS

Many types of specialty work require the electrical contractor to work directly with a manufacturer of specialty equipment. By dealing directly with the contractor, the manufacturer can make sure that the system is properly installed and serviced. The manufacturer usually trains the contractor and his people to install the system. In return for this, the manufacturer expects that the contractor will install only his brand of products. For this reason, most specialty-system installers install only one brand of product, those of the manufacturer that started them in the business. This is particularly true for energy-management systems, fire-alarm systems, lightning-protection systems, programmable controllers, and security systems.

By virtue of this relationship with the manufacturer, the contractor becomes a sales rep as well as the installer. Because their materials don't follow the channels that most electrical material follows, these specialty manufacturers do not have the benefit of manufacturer's reps or distributor salespeople promoting their products. They therefore rely upon the contractor to do most of the sales work. This requires the contractor to do things that are not typical of the electrical contracting business, such as calling on architects and engineers, putting out a number of written quotes, and making sales calls on customers. Because of this, many electrical contractors who have gone seriously into specialty contracting have needed to set up a separate company to do only the specialty work.

If the volume of work is available in your local area, specialty work can do very well for you. The following sections discuss specific specialties.

Energy management. In the mid-seventies it seemed that energy management was going to be the wave of the future. Oil prices skyrocketed, taking almost every other type of energy cost along with them. There was a great interest in different types of energy management. But after a number of years, the price of energy dropped, and with it the interest in energy management. So energy management has taken somewhat of a roller-coaster ride, but the fact remains that there is a real need for energy management. The future prospects for the EM (Energy Management) market are good. While the price of energy has recently been low, there is little question that in the long run the price of energy will go up again, maybe dramatically. Once prices go up, it won't take long for people's interest in EM to pick up again. The question of growth in the energy-management market is more of *when* than *if*.

Programmable controllers. Programmable controllers (and robotics) are here to stay in the industrial market. The amount of time they can save and the amount of control they can give to management are very valuable to all but the smallest industries.

Virtually any factory is a customer for some type of process control, but to sell your services well, you need to first establish credibility.

If you decide to get into the PC (Programmable Controller) market, be very careful in choosing the manufacturer that you will represent. This manufacturer should offer the products that your specific customers need, and be able to help you (and your customers) both before and after the sale.

Lightning protection. The lightning-protection market is greatly determined by geography. There is quite a bit of lightning-protection work available in the southeastern states, but much less in the western states.

Where enough work is available, it can be very profitable, although there are very few markets in the United States that can support a full-time lightning-protection contractor. There are a number of qualified manufacturers in the market, and the Lightning Protection Code (NFPA 78) is very easy to learn. Licensing is also obtained without any great difficulty.

Fire-alarm work. Fire-alarm work is one of the more plentiful types of specialty work, since so many buildings (typically public buildings) are required to have some type of fire-alarm system.

Fire-alarm work will require you to work very closely with the manufacturer, and will almost require you to set up a separate company, as other electrical contractors will be your customers.

The manufacturers are usually more than willing to provide training, provided that there is a substantial market available and that the contractor is willing to make a commitment to selling the manufacturer's products. Again, the size of the market determines the viability.

Telephone work and cable TV. Telephone work and cable TV are the two biggest specialty markets that have opened up in quite some time. The breakup of AT&T has spawned a whole new type of contracting, the telephone-interconnect market, and cable TV has been rapidly growing for a number of years.

For these markets, a contractor is almost always better off to set up a separate company, but it has been well worth it to the contractors who have done a good job.

Where these markets will go in the future is still unclear, although it is difficult to imagine that either of these markets is going to diminish by any great percentage.

Other markets. Whenever a special need arises, a specialty market is born. In recent years there has been a need for expert services in PCB removal, contractors who know how to set up computer rooms, and wherever there are road projects, contractors who can do traffic lights.

The various markets are developing all the time, but it takes a contractor who keeps his eyes open to cash in on them. Keep your eyes *and* ears open, and always think things over several times before you commit your company to any new venture, but if you see a real opportunity don't be afraid to take it. When the opportunity is new, the demand is high and the supply is low; this means better profits for your company.

8

DESIGN/BUILD WORK

Design/build work is different from regular electrical construction in that the electrical contractor plays the role of both the contractor and engineer, hence the name Design/Build.

Design/build work has several inherent advantages over contract-bid work:

1. The amount of competition for jobs is far less, sometimes nonexistent.
2. The electrical contractor is compensated for providing the design services that are usually done by an engineer. (This compensation usually comes in the form of a higher percentage of markup.)
3. The customer usually gets a better price for the same quality of work by using a qualified design/build contractor.
4. The design/build forum gives the contractor a better opportunity to negotiate an acceptable package of price and services.
5. Cooperation during the construction process is usually better on design/build projects than on contract-bid projects.
6. Design/build work usually carries less of a risk than many other types of electrical contracting.

The advantages of design/build work are many, but there are several important prerequisites to selling and performing design/build projects:

1. The contractor must develop a great deal of trust in his customers.
2. The contractor must offer the customer a better value for the money than is available on the contract-bid market.
3. The installation must be done in a first-class manner at all times. Design/build is a premium type of service; if it is degraded to something less by shoddy installation techniques, its value drops dramatically.
4. The contractor and all of his personnel must maintain a high standard of courtesy to the customer.
5. The customer's interests must take precedence over the contractor's interests.

GETTING DESIGN/BUILD WORK

There are two primary sources for obtaining design/build work. The first is through general contractors who do design/build projects and need a qualified electrical contractor to handle the electrical-systems installation.

The second is by selling directly to a customer who needs an installation that is all (or almost all) electrical.

General contractors have certain considerations in choosing an electrical contractor for a design/build project. This choice is important to them, because they are choosing a partner to work with throughout the entire project. When the project runs smoothly, the general contractor makes money, but if the job is plagued with problems, the general contractor loses profits. So the choice of subcontractors is very important to a general contractor and, as opposed to contract-bid work, in design/build work they have a wide-open choice. The general contractor's main criteria for choosing an electrical contractor for design/build work are typically these:

1. The electrical contractor's ability to properly design the project. This means that the project must meet all codes, and also meet all of the customer's needs, while being done for a reasonable cost.
2. The electrical contractor's ability to perform the installation properly. First of all is the requirement to do the actual installation correctly. But along with this is the requirement of the contractor to work with the general contractor and his field personnel. In design/build work, the general contractor and the electrical contractor are partners.
3. The electrical contractor's ability to handle all of the details of design and coordination with utility companies and municipalities.

Figures 8-1 and 8-2 show proposal letters for design/build projects. Notice that each letter specifies a lot of exact quantities. This way, if there is a question of costs, each individual part of the project can be evaluated independently and decisions can be made as to where costs can be trimmed and where they cannot. For design/build work, the proposal is important, as it should convey your company's professionalism directly to the customer.

Selling directly to the customer is particularly good for larger modernization projects or installations of new types of systems. This includes computer rooms, programmable-controller installations and other types of special process controls, and lighting retrofits. The same type of proposal letter is used for selling directly to the customer, and the list of considerations are basically the same.

Why design/build can be a better value. If it is properly done, design/build is usually a better value for the customer. (The emphasis is on the *if*. Not every contractor who says he can do design/build really does it properly.) These are several of the reasons why design/build can be a better value than the standard contract-bid arrangement:

February 18, 1987

Mr. Bill Mitchell
Wade Construction, Inc.
2147 Gordon Terrace
West Orange, NJ 08421

Dear Bill,

Per our conversations of the past two weeks, the following is an itemized preliminary estimate for the Senior Citizen Housing project in Marshall, NJ.

The wiring in the individual units will be type NM cable (copper), according to the N.E.C. and good trade practices.

The electrical service to each unit will be 100 amp, single phase, 120/240 volt; distributed through 34 main lug-meter centers. The meter centers are to be supplied by approximately 7 pad-mounted transformers in various locations. We will provide feeds from the transformers to the meter centers and the transformer pads. The local utility is to be responsible for the transformers, connections, and all high voltage primaries.

Special systems to include:

A fire alarm system with an integral voice evac (PA) system; a nurse call system, with a main station in each unit, and a remote station in each bathroom; telephone raceways as required, service to units to be by utility company.

It should be noted that all special systems conform fully to the NFPA and HUD requirements, to form a complete, reliable system. All control panels to be located in the common facilities.

The itemization is as follows:

128 Type A Units	$107,000
50 Type B Units	33,400
12 Type C Units	12,200
10 Type D Units	5,000
Common Facilities 13,500 sq. ft.	81,000
Light Fixtures for 200 Units	30,000
7 Transformer Pads, etc.	4,200
33 6 Gang Meter Banks	55,100
1 2 Gang Meter Bank	800
Feeds, Transformers to Meter Banks (4,770 Ft.)	62,400

Figure 8–1

Feeds, Meter Banks to Units (11,200 Ft.)	38,500
Special Systems 200 Units	90,000
Preliminary Total	$519,600

This estimate was prepared from plans dated 12/7/86 (4 pages) prepared by Philip Webb Associates, and William Hurston Architect, and represents a complete, organized and professional installation.

We look forward to working with you.

Sincerely,

James McAltus
Chief Estimator

Mike Hardin, Const. Manager
Forster Properties
1744 Brook St.
Barnes, GA 33120 10/18/87

Dear Sirs,

The following is an itemized estimate for the "Deer Run Condominiums" project in Auburn, GA.

The wiring in the units will be type NMB cable, according to the N.E.C. and good trade practice.

The electrical service to each unit will be 150 amp, single phase, 120/240 volt; distributed through multi-gang meters located at each individual building. The local utility is to furnish the meters, meter bases, and all primary feeders to the meters. The service to each unit will be in conduit below the floor slab.

The addition to the items expressly shown on the electrical drawings, this estimate includes the following:

Wiring for a dishwasher in each unit.
One telephone outlet in each kitchen, living room, and master bedroom.

Figure 8–2

One TV outlet in each living room and master bedroom.
Telephone and TV service cables from the service point to the first outlet will be in conduit; providing an independent service to each unit.
One range hood in each unit (wiring only).
We will furnish all exhaust fans shown (all required ducting by others).
One doorbell chime and pushbutton in each unit.
One 30 amp circuit and disconnect for AC unit in each unit.
One 50 amp circuit and required disconnect for each air handling unit.
2 smoke detectors in each unit A,B,D,E. One in each type F or G unit.
All outdoor and bathroom receptacles to be GFI protected.
Each unit to have 1 outdoor receptable.

All light fixtures are to be furnished by others and installed by this contractor. No temporary power is included in this estimate.

The prices for each unit are as follows:

Unit A	$1,670.00
Unit B	1,531.00
Unit D	1,571.00
Unit E	1,630.00
Unit F	1,584.00
Unit G	1,545.00

This estimate was prepared from plans dated 9/17/87 as noted by Forster Properties, and represents a complete, organized, and professional installation.

We look forward to working with you.

Sincerely,

Nolan Harper, VP

1. The electrical contractor is more intimately aware of different types of electrical materials and installation techniques (and their respective costs) than is an engineer. Design/build work puts the burden of specifying upon the contractor rather than upon the engineer. Since the contractor is usually more familiar with which types of in-

stallations are less expensive, it usually makes for a more economical job.

2. Some engineers (by no means all) have a tendency to gold-plate their jobs; that is, they tend to use unnecessarily expensive types of products on their jobs. There are two reasons that engineers have for doing this. The first is that they like all of their jobs to be first class, a type of professional pride. The second is that since engineers get paid by a percentage of the contract amount, the more the job costs, the more the engineer gets paid. (Ideally this should never happen, but sometimes it does, whether it be consciously or unconsciously.)
3. Because the electrical contractor does the design work, the contractor's people will be far more intimately familiar with the project than if they had simply bid the job. This tends to make the entire installation process more efficient. Questions can be quickly and easily answered by the person who actually designed the installation.

DESIGN OPTIONS

The following is a list of various design options to make electrical construction projects more economical without degrading the quality of the installation:

1. Light fixtures should be carefully chosen to get the best lighting at the lowest cost. Choose fixtures that will give the lighting and look that the owner wants, but no more. Try to use as few types as possible; this makes the installation go a lot smoother with less confusion.
2. Use die-cast, set-screw EMT fittings through 1¼ inch. Larger than this, go to a compression-type fitting, as some die-cast set-screw fittings in the larger sizes will crack when over-tightened. Set-screw fittings are quite a bit easier to install than compression fittings, and die-cast fittings are about half the cost of steel fittings and do virtually the same job.
3. Where sound-proofing isn't an important consideration, use as many back-to-back receptacles as possible. This eliminates an average of 7 feet of conduit and 20 feet of wire every time you can do it. The amount of sound that can actually be transferred through these receptacles is actually very small, but in some situations you may need to avoid it.
4. Use no more than one home run for every two or three circuits except when necessary. There is nothing in any way wrong with putting two or three circuits in the same conduit, provided that they are all from different legs (or phases) and fairly well balanced. Obviously the conduit must also be properly sized.

5. Use PVC conduit for underground and underslab runs whenever possible. PVC is quite a bit cheaper and easier to install than GRC conduit, and will not corrode like GRC. However, you should use steel elbows in these runs (GRC or IMC conduit elbows). This is because the elbows get a lot of pressure and wear when pulling wire; a pull string can actually cut through a PVC elbow. These GRC or IMC elbows should be thoroughly coated with bitumastic paint before they are put in the ground to protect them from corrosion.
6. Using aluminum wire for services, feeders, and large circuits is quite a bit less expensive than using copper wire, but you should make sure to check the connections after a couple of months of service under full load, as aluminum connections loosen more than copper connections (although some of the new aluminum alloys do very well).
7. For most purposes a residential grade or medium grade of wiring device will do just as good a job as a spec-grade device (at a fraction of the cost).
8. Be careful how you use THHN, XHHW, and other 90-degree centigrade-rated wire. The ampacity figures look wonderful, but they are based upon a temperature of 90 degrees centigrade. The problem is that most circuit breakers and lugs are only rated to 75 degrees centigrade; thus making them unsuitable for the higher temperature. There has been a certain amount of confusion between the wire manufacturers, the lug manufacturers, contractors, and inspectors over this matter. Check with your local inspector before you find yourself in a jam.
9. Don't size your conduits too tightly. If you fill them as full as code allows, you will probably waste a lot of money by increasing the difficulty of pulling the wire. Beyond the initial loss of time, the morale of the workers tends to drop when performing an unnecessarily difficult task, making productivity even worse.
10. Provided you have electricians that can do it properly, run as much of your power circuitry under the floor slab as you can (see Chapter Five for details).
11. Using loadcenters instead of panelboards is fine for light duty uses, but where there will be heavy use go to a bolt-on type panelboard.
12. General-duty safety switches will do the same job as heavy duty in all but the more demanding applications.
13. Low-voltage switching has not been in vogue for a good number of years, but it gives the customer a great deal of flexibility in switching and control and is usually quite economical.
14. A well designed Manufactured Wiring System (see article 604 of the

NEC) can save a great amount of time on a recessed lighting installation, *but* make sure that the fixture modules are installed at the factory whenever possible.

One of the most major factors in the design of electrical systems is that the entire installation should be designed for making a profit, rather than being designed to meet code only. Certainly code must be met, but there is a lot beyond that which can be done to make a job more profitable. Design/build work gives the electrical contractor the opportunity to design the job as best suits him. So every design decision should be based on two basic criteria: First, the design must meet code, and second, is it the most profitable way to do the job?

Performing the Installation

As in any electrical installation, it is critical that the electricians get the best possible instructions for putting the job together. In the case of design/build work, the contractor has even more than the usual opportunity to make up several good sets of working drawings. All types of job planning become easier when the electrical contractor also does the design of the electrical system. When you do design/build jobs, remind yourself of the importance of these things, and take as much in profits out of the job as possible.

Other notes. In the design/build process things often move very slowly, so be prepared for some delays. The design process is not as fast moving as most electrical contractors would like, and it often takes a while to get answers out of municipalities and utility companies.

As a general rule, do not give a customer a set of blueprints that you designed without getting a signed contract. (Actually you should get some money, but a signed contract is the minimum you should accept.)

When you submit a proposal, give the customer several options with the cost of each clearly stipulated. This gives you a starting point for negotiations.

Pay great attention to the customer's needs during all of the design process. With design/build, you are selling the customer complete satisfaction. Pay attention to the needs both present and future; go the extra mile to be sure that the customer's needs will be well covered.

Look after your customer's interests. Saving the customer money is what will sell you future jobs. The customer is trusting you as an expert to take care of his needs; do everything that you can.

9

CONTRACT-BID WORK

For years contract-bid work has been the mainstay of many electrical contractors. There has been a reasonably sufficient amount of work in most markets to keep the electrical construction business in gear, and therefore the markets have gotten the best of most electrical contractor's efforts. Over the years competition has become increasingly fierce; yet the work has been fairly consistent.

The profit margins for contract-bid work are usually the smallest in the business, but the amount of available work is the largest. This makes contract-bid work inviting to the contractor who would like to do a high volume of work and doesn't mind doing it at a lower percentage of profit. Doing contract-bid work at a reasonable percentage of profit requires the contractor to work a lot smarter and more efficiently.

GENERAL CONTRACTORS

Because the general contractor is the electrical contractor's boss, dealing with the general contractor is extremely important. The electrical contractor must understand how a general contractor operates and what things are important to the general contractor.

A general contractor is an organizer and coordinator. On any of their construction projects, general contractors have to oversee the work of a large number of different trades. They have to be very thorough, and make very sure that the proper work is done by each of the various trades. If any one trade falls behind on its work, the entire job will suffer. To make money on a construction project, the general contractor has to be sure that everything runs smoothly. The more hassles and problems that a job has the less money the general contractor makes on the job. General contractors make their money by finishing jobs ahead of schedule with a minimum of job problems.

As in any industry, there are some general contractors who do a better job of running projects than others. Because of this, it is important for the electrical contractor to consider each general contractor's characteristics before submitting a bid. Some general contractors do a first-rate job of project management, which makes the job a lot easier for the electrical contractor. A GC (General Contractor) like this should get a slightly lower price on any bid than a different GC who does a poor job of job management. Most electrical contractors give very little consideration to the general contractor when submitting bids, which is unfortunate. The way the GC runs a project is one of the most important factors in the profitability of the project. Submitting a bare-bones price to a GC who is going to do a lax job of running the project can make the job end up a loss for the electrical contractor. A GC that runs jobs well should get a lower price, and a GC who doesn't run jobs well should get a higher price. It is far more important than most electrical contractors realize to understand the general contractors with whom they work. You should know each one's strengths and weaknesses, even the personal traits of the key people at all of the local general contracting companies.

On the job site, it is very important that your job foreman stays in close touch with the GC's superintendent. No one else can affect the job as much as this person can. Your foreman should have daily contact with the superintendent. They should discuss the flow of the job, which other trades are either holding up or pushing your work, and any other items of interest relating to the project. Keep this person happy with your company; as a friend, he can help you a lot, but if things go the other way, this man can be one of the worst enemies your company can have.

Bonding

For contract-bid work, bonding can be critical. In return for a fee (usually around 1.2% of the contract price), a bonding company will guarantee the project's completion to the owner or GC for the specified amount. Many projects require this for obvious reasons. So, the ability to obtain bonding when needed can be of great importance. Bonding companies are careful, as you might expect. They like to see financial stability and expertise before they will bond an electrical contractor. In addition to this, they set a limit as to how much bonding they will give to that contractor for the year; once the contractor has reached the pre-specified limit, the bonding company will bond no more of the electrical contractor's jobs until some of their work is completed. For these reasons it behooves the electrical contractor to make every effort to impress the bonding company's agent. These agents will stop in to see the contractor's office and jobs. Their decisions on bonding are based first upon the financial picture of the company, but after that, the agent's impressions come into play. There is a human factor to this decision-making process, not just the mathematical factor.

Contracts

Since this type of work is done exclusively by contract, the contract itself becomes very important. The content of these contracts is extremely important; the contract spells out exactly how the job will be run, and how your company will be paid.

The most common type of contract in use is the AIA contract. This contract, while not perfect, is considered a fair compromise for all involved. Some of the contractor's associations have their own contracts, and some of those are pretty good also. But remember that these are contract forms;

any notes made to these contracts can make them very dangerous.

Take care as to what types of contracts your company signs. If the wording is unclear, take the contract to your company's attorney to see what the attorney thinks of the contract and what changes she thinks should be made to protect your company. Remember that these contracts are legally binding documents, and that one poorly worded phrase can put your company into bankruptcy.

Do not ever be afraid to make any reasonable changes to contracts. If there is some part of the contract that is unfair to your company, do not be hesitant to ask for it to be changed. If your requests are reasonable, the other party should not have any problem with the changes. However, if the other party will not agree to reasonable changes, you have good reason to believe that they are not going to be reasonable during the course of the project either. The stakes are too high to just go with what is easy; you need to look after your company's interests, and almost nothing is more important than the wording of the contract. You are better off to lose a project than to lose your company. (It is also worthy of mention that the other party to the contract will be put in a much less favorable position if they can't come to an agreement with you on the contract. If they can't agree, they will have to go back to the other electrical contractors [who know that your company was the low bidder] and try to renegotiate with them to do the job. More often than not, these other electrical contractors hesitate to do this, as they will wonder why your company turned down the job.)

Be careful of contracts that are drawn up by the individual general contractor. Very often these contracts are strongly tilted in the general contractor's direction.

LEGAL CONSIDERATIONS

In contract-bid work, more than almost any other type of contracting, having good legal advice is very important. It is almost mandatory for your company to have a lawyer who is very familiar with construction law. Like any other profession, lawyers have specialties; find a lawyer whose specialty is construction law. You can find these firms by talking to other contractors in your area, both electrical contractors and other types of construction contractors. The top people in your company should meet with the lawyer at least once per year, and discuss any and all legal questions facing your company.

You should have a thorough understanding of your state's Mechanic's Lien laws. Every state is different, so it is important to talk to a lawyer who is familiar with construction law in your state to find out the particulars for your area. In many locations the contractor must file a notice to the project owner in order to protect all of his lien rights. If this (or anything similar) is the case in your area, you should never fail to do this. Even if the people with whom you are dealing seem as good as gold, you don't know what tomorrow will bring; there are just too many things that can go wrong to leave yourself open. The people with whom you are dealing may be great, but who is to say that someone might not do something to ruin their company, so that as good as their intentions are, they become unable to pay you.

Documentation

In any type of contracting it is very important to get anything of importance in writing. Oral communications are virtually impossible to prove in court; written communications stand up far more. Documentation of job delays, change orders, early acceptances and the like are very important.

As far as job delays are concerned, be especially careful on projects where the liquidated damages are specified by the contract. A liquidated-damages clause in a contract specifies that a certain amount of money will be paid to the owner (or deducted from the amount due the contractor) for every day that the completion of the project extends beyond the specified completion date. These charges are frequently $200–400 per day and higher. On a job with any substantial amount of delay, this can run into a large amount of money quickly. Take every precaution to protect your company from these type of charges by strongly documenting any delays caused by others that affect the progress of your company's work (See Figure 3–7).

Arbitration

Arbitration is a method of resolving contract disputes without going through the regular legal channels. Arbitration is a simple process that is voluntarily chosen by both parties to resolve a dispute. The dispute is resolved by an impartial judge mutually selected by both of the parties. Both parties agree in advance that the judge's decision will be binding and final. As a rule, courts will not overrule the arbitration judge's decision unless it can be proven that the judge was partial to one of the parties (which almost never happens).

The advantage of arbitration is that it is much quicker and cheaper than resolving a dispute in the courts. Because of this, arbitration is specified for job disputes by many construction contracts. The most frequently used arbitration procedures used are those of the AAA (American Arbitration Association). The AAA rules for the settlement of disputes in the construction industry are included in the standard AIA (American Institute of Architects) contract form.

The bottom line of the comparison between arbitration and normal litigation is this: From the time that the first demand for arbitration is filed until the final decision is given is about six months, while litigation is almost never completed in less than a year. The initial expense of an arbitration request is only $50, and the total expense for an average claim under $10,000 is only about $400, which is a fraction of the cost of litigation.

BIDDING

The process of bidding on construction projects is one of the most unrefined of any industry. It is continually subject to deception and cheating. At times it can be a very scientific process, while at other times it can be one big tumult.

Figure 9–1 shows a pre-bid checklist. This form contains all of the important information on the project, so that it can be quickly reviewed before the project is bid (or sometimes before it is taken off). The top sections are for the basic facts of the project; name, location, bid date, building construction, financial information, and so on. The middle section lists the special quotes that will be required to bid the project and specifications for the various types of material which will go into the project. The bottom sections are to show the cost responsibility for different parts of the job and for the list of bidders and any notes. In the cost-responsibility section there are several abbreviations shown. FBE means furnished by electrical contractor; IBE means installed by electrical contractor; FBO means furnished by others; and IBO means installed by others.

Figure 9–2 shows a simple quote sheet for writing down any quotations received from suppliers over the telephone on the day of the bid. All of the information shown should be filled in, including the date and time the quote was received.

Figure 9–3 shows a simple bid summary, where the final bid tabulations are done. Any quotes will be filled in, along with the costs of the basic wiring materials; labor hours and job costs will be filled in; a certain percentage of the total cost will be added for overhead and profit, and the job will be ready to bid. The simplicity of this form is important. Since the general rule is that the final bid tabulations are done in a big hurry just a few minutes before the deadline, a complex bid summary can make for confusion and mistakes. A simple bid summary such as the one shown makes the entire process easy to do and revise should you get a late price from a supplier.

Figure 9–4 shows a list of bidders. This list should be filled in as the bids are called in to the general contractors. The section at the bottom is for any exclusions or details that need to be communicated to the

Pre-Bid Information

BIDDING INFORMATION
Name of job ______
Location ______
Bid Date ______
Bids to ______

DESIGN INFORMATION
Architect ______
Engineer ______
Owner ______

FINANCIAL INFORMATION
Bid Bond ______
Construction Time ______
Liquidated Damages ______
Retainage ______
Wage Rate ______

CONSTRUCTION INFORMATION
Building ______
Floors ______
Walls ______
Ceilings ______
Elevations ______

QUOTES REQUIRED
Lighting Fixtures ______
Distribution Apparatus ______
Fire Alarm ______
Clock ______
TV ______
Sound ______
Intercom ______
Nurse Call ______
Doctor's Register ______
Security ______
Lightning Protection ______
Electric Heat ______
Snow Melting ______
Heat Tracing ______
Emerg. Generators ______
Dimming ______
Stage Lighting ______
PCs ______
Energy Management ______
Signal ______
Isolating Panels ______
Cable Tray ______
Flexible Wiring ______
U.F. Duct ______

SPECIFICATIONS
Devices
- Switches ______
- Receptacles ______
- Wall Plates ______
- Special Outlets ______

Branch Wire
- Type ______
- Separate Ground ______

Feeder Wire
- Type ______
- Separate Ground ______
- High Voltage ______

Raceways
- Underground ______
- Slab ______
- Surface ______
- Partitions ______
- Fittings ______
- Minimum Size ______
- Supports ______
- Boxes ______

COST RESPONSIBILITY	F.B.E.	I.B.E.	F.B.O.	I.B.O.
Temporary Electrical				
Excavation & Backfill				
Concrete Encasement				
Equipment Pads				
Pole Bases				
Fixture Support				
Cutting & Patching				
Pitchpockets				
Sleeves				
Painting				
Starters				
Disconnects				
Control Wiring				
Interlock Wiring				
Primary Transformer				
Primary Raceway				
Primary Cable				

BIDDERS

______

NOTES

______

Figure 9–1

Quote Sheet

Job: ____ Date: ____________________

Supplier	Item	Price	Bill of Materials	Date & Time

Figure 9-2

Bid Summary

Project:________________________ Bid Date:____________

Misc. Material

Fixtures

Gear

Material Cost____________

Sales Tax____________

Total Materials____________

______hours @ ____ . ___ = ____________

Job Costs-(__________) - ____________

Net Cost-____________

% O & P-____________

Total-____________

Bid @ — $____________

Figure 9–3

general contractor. By writing these details down, they will not be forgotten, which could be disastrous.

If your company should ever make a mistake in one of its bids, be sure to correct it immediately, both orally and in writing. By correcting it immediately you can get released from it most of the time; but if you let it go any more than that, your errant quote becomes legally binding.

Always be sure that your bids include expenses for as-built drawings, maintenance instructions, testing, and the like. These items are easy to forget or ignore during the high-pressure bidding process, but you will have to pay for them later.

If your company is low bidder on a certain project, but it will take some time for the contract to be written and signed, a general contractor will often send you a letter of intent. A letter of intent states that the general contractor has been chosen to receive the contract for the construction of the particular project, and that he intends to use your company as the electrical sub-contractor. This letter gives you some legal right to the job,

List of Bidders

Project: ______________________________

Bid Date & Time: ______________________________

Name	Phone	Date & Time	Person's Name	Base Bid	Alt.	Alt.	Alt.	Alt.

Exclusions, etc.:

Figure 9–4

as it is a statement from the general contractor to your company that the general contractor will give your company a contract on the project as soon as contract from the owner is received.

After you are in possession of a letter of intent, your company can proceed to finalize its pre-job planning, and notify your suppliers of your intent to buy from them. This is important, because if there is a substantial delay between the bid date and the time when you get the contract, your suppliers may not be able to guarantee you the same price as you used in the preparation of the bid. With the letter of intent, you can finalize the terms of your material purchases. You can even write the purchase orders, adding a clause stating that the purchase is contingent upon your company being awarded the contract. You can then begin preparing submittals and the like.

CHANGE ORDERS

Change orders can be one of the biggest problems in a contract-bid project. In general, owners would like to see no change orders whatsoever, while contractors would like to have plenty. A change order is an order (usually written) to a contractor from his customer (typically a general contractor) to add or delete specified work from the parties' original contract.

Change orders are a hassle and unexpected expense for the project owner. They represent money that the owner will have to spend, but wasn't really planning to spend. Contractors tend to like change orders because they usually make a higher percentage of profit on change orders than they do on the contract work. (Some electrical contractors will even take a job for just above cost, hoping to make money on expected change orders; obviously this can be very dangerous.)

Many owners (and architects) feel that contractors gouge them on change orders, and often rightfully so. Because of this they have included clauses in their contracts or specifications that specify certain percentages of overhead and profit that can be charged for change orders. Usually these percentages are ridiculously low, well below the contractor's actual overhead expenses. But regardless of this hard-line wording, the contractors still are not restrained as the owners would like. This is because the contractor is almost always asked to provide an estimated cost for the change order before authorization is given. The contractors overestimate the cost, and then add the ridiculously low percentages of overhead and profit to the inflated price. Once the work has been approved and completed, the contractor submits a bill for the agreed-upon amount.

It is prudent to get all change orders in writing. Some general contractors and owners are notorious for not paying for change orders, even if the work has been done on a verbal authorization. If the order is in writing there is very little question that payment is due to the contractor. There is no good reason that a general contractor or owner should not be willing to give written authorization for a change order. If they are willing to pay for the work, they should have no problem with signing a letter to that intent before they expect you to send workers and materials out to the job.

10

RESIDENTIAL WORK

Residential electrical work is unique in that it allows the electrical contractor to come as close to the assembly-line type of production as electrical installations will permit. Indeed, mobile homes are wired assembly-line style, and mobile homes are wired almost the same way as residences.

The real profits in residential wiring go to contractors who work exclusively in wiring residential dwellings or who have a somewhat autonomous branch that works exclusively in the residential market. By doing this, the contractor is able to specialize his company's operations for doing residential wiring. The company has all of the necessary tools and equipment for doing residential wiring, and has no need to stay geared up for commercial or industrial wiring. This simplifies scheduling, reduces the amount of inventory, and makes it easier to train electricians.

To specialize so completely in one type of work it is essential that there is a large enough market to keep the contractor busy. While this isn't so in every market, residential wiring is probably the best suited sector of the market for this. Home building is very cyclical, but the figures are measured in millions, which means that even in slower times there are a lot of houses being built; but again, there are some areas which will not sustain this type of operation.

PRICING AND PROFITS

Price levels (and thereby profits) in residential wiring are determined by the marketplace, but in most areas residential work still allows a much higher percentage of profits than most other types of work. The reason for this is probably that each project is fairly small (with the exception of multifamily housing), and that most contractors (especially large contractors) charge more for smaller jobs since they have to set up each one individually.

The profits in residential work come from *getting in* and *getting out*; that is, by doing the jobs very quickly and thereby saving on labor expense. With workers who are properly supplied, trained, and supervised a home can be wired in about twenty hours per thousand square feet, sometimes even less. This is obviously very fast, but attainable if the company's operations are specifically set up for residential work.

The best and fastest method for pricing residential work is by the unit-pricing method. The idea of unit pricing is simple: to establish a set dollar amount for each item or group of items that is identified by one symbol or name. For example, a customer may ask for the price of a receptacle. However, the customer doesn't want a price for the receptacle alone, but for the receptacle, box, wire or cable, trim plate, and labor. He wants a total price for each item that he can see and use, and he identifies it by one name, a receptacle. With a good set of unit prices (unit prices for all of the commonly used items), the contractor can price a job by simply counting the different symbols on a set of electrical plans and multiplying them by their unit prices to determine a final selling price. The entire process for an average residence can be completed in about ten minutes. For specific information on developing a set of unit prices refer to RESIDENTIAL AND COMMERCIAL UNIT PRICING FOR ELECTRICAL CONTRACTORS, which is published by Prentice Hall, Englewood Cliffs, NJ 07632.

For an average house, the material cost runs between 50 and 60 cents per square foot (1987 figures; fixtures and fans not included), and labor runs at between 19 and 24 hours per thousand square feet. These figures assume a reasonable amount of work.

Take care when doing custom homes that you don't let yourself get run-around changing things. You can understand how someone who is building a custom home would want to have everything perfect; after all, they are spending maybe $100,000 or more for this house and want it to be perfect. You should allow extra time (and charge for it) to take the owner around the house and make sure that everything is the way he wants it. It is best to talk to the owner a day or two before you rough-in the house, and tell him something like this: "Sir, if you have any changes that you would like to make, we need to make them by tomorrow morning. Tomorrow morning we will be starting the actual installation of the wiring, and after we start, any changes are going to cost you money. Any changes in locations that you make now will be at no charge; but if we have to change or move something after tomorrow morning, I am going to have to charge you for it. So please, take your time, ask any questions that you have, and let's resolve everything right now while it is easy to do." Doing this will save you a lot of grief on custom house installations.

TOOLS AND MATERIALS

The types of tools and materials that are used in residential work are fewer than those used by full service electrical contractors and electricians, which makes operations a lot simpler and less expensive. The tools that your company's electricians should carry are the following:

Hammer (20-ounce)
Lineman's pliers (9-inch)
"T type" cable stripper
Blade screwdriver (large)

Blade screwdriver (small)
Phillips screwdriver (large)
Phillips screwdriver (small)
Channell lock pliers (10-inch)
Hex wrench set
Razor knife (retractable)
Hacksaw frame
Voltage tester ("Wiggy" type)
Flashlight with continuity tester attachment
Carpenter's apron
Tape measure (1″ x 25′)
Pencils and markers
Long-nose pliers
"T type" wire stripper

Beyond the tools that each electrician is expected to carry, the contractor is responsible for providing other tools that will be required to perform the installations. These types of tools include:

Right-angle drill
⅜″ variable-speed reversing drill
Circular saw
Drill bits:
 1″ auger bit
 2″ auger bit
 ¼″ masonry bit
 Small set of high speed steel bits
Extension cords
6′ step ladder
20′ extension ladder
Medium-sized shovel
Post-hole digger
Fish tape
Stilts
Cable dishes
Hacksaw blades
Wire lube
PVC glue
Electrical tape

You will notice that stilts and cable dishes are listed. These are somewhat unusual types of tools for electrical contractors to use, but they are tremendous time savers if used properly.

Figure 10–1 shows a cable dish. This dish is specifically designed to be used with type NMB cable, commonly known as Romex. This type of cable comes in either boxes of 250 feet each or on 1000-foot reels. The boxes are awkward to use because the cable is difficult to remove and gets twisted when it is pulled out of the box. The reels are difficult because they are very heavy and awkward to move around the job. The cable dish makes it possible to use the cable easily.

The cable dish is made from a circular piece of plywood approximately 18–22 inches in diameter. An eye screw is put into the center of the plywood. To the eye screw is attached a swivel snap (just like the end of a dog's leash), and to that about 3 feet of chain with a large hook on the end.

By removing a 250-foot coil of cable from the box that it comes in and placing it on the cable dish, then hanging the cable dish from one of the ceiling joists, the dish will dispense the cable smoothly and evenly. It helps to put a little bit of oil or grease on the swivel snap, and to put a piece of scrap PVC conduit over the bottom 6–9 inches of chain, so that the cable won't get torn as it brushes up against the chain.

Using stilts for installing cable works very well because it eliminates almost all of the time that the electrician spends climbing up and down a ladder. The type of stilts to use are the same ones that drywall installers use. They can be purchased from almost any good contractor supply company for about $250. The stilts should be set up so that the electrician's head is just below the bottom of the ceiling joists. This way he can almost

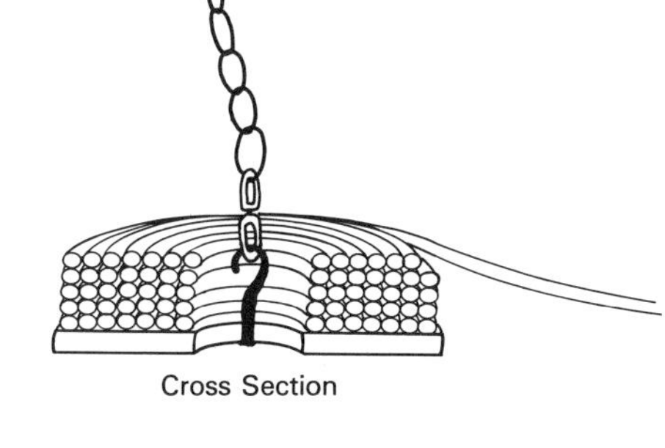

Figure 10–1 Cross Section

run through the project installing cable once he gets the system coordinated.

There is one important thing to remember when using stilts: All of the drilling and notching of the structure's framing must be done before the electrician gets up on the stilts. If the electrician has to drill holes while on the stilts, the time saved by using the stilts will soon evaporate. Also be sure that the floor of the building is fairly clean, as you don't want anyone to fall off of these stilts.

Electricians who are learning to use stilts should have someone stand right by them for the first several times until they are sure-footed on the stilts. Once the stilts are mastered they can be used very safely, but care must be taken while the worker is learning to use the stilts: A fall from two feet off the ground hurts a lot more than a normal fall.

The number of different types of materials used in residential wiring is a lot more limited than for other types of electrical installations. For interior work the following list of materials will cover 99 percent of the materials that will be required:

1 gang nail-on boxes
2 gg (gang) nail-on boxes
1 gg flanged boxes
2 gg flanged boxes
3 gg flanged boxes
4 gg flanged boxes
Round nail-on boxes
Pancake boxes
½-inch, 1 gg plaster rings
Duplex receptacles
GFI receptacles
Range receptacles (surface or flush)
Dryer receptacles
Smoke detectors
Door chimes and push buttons
Single-pole switches
Two-pole switches
Three-way switches
Four-way switches
Dimmer switches (600 watt)
60-amp non-fused disconnects

30-amp non-fused disconnects
1 gg receptacle plates (plastic)
2 gg receptacle plates
1 gg switch plates
2 gg switch plates
3 gg switch plates
4 gg switch plates
Dryer-receptacle plates
Range-receptacle plates (for flush receptables only)
Weather-proof receptacle plates
Weather-proof switch plates
Blank plates
TV plates
Phone jacks
Combo switch and receptacle plates
½-inch Romex connectors
¾-inch Romex connectors
Staples (large)
Staples (small)
"F type" clamps (also called Madison hangers)
#14/2 cable
#14/3 cable
#12/2 cable
#12/3 cable
#10/2 cable
#10/3 cable
#8/3 cable
#6/2 cable
#6/3 cable
Phone cable
Coaxial cable (TV cable)
½-inch Sealtite and fittings (non-metallic)
#10, #8, & #6 THHN wire (short pieces for equipment connections)
Nails
Sheet metal screws
Plastic anchor kits
Masonry nails
Toggle bolts

Misc. washers (especially ¼" x 1¼" fender washers)
UF cable (#14/2 or #12/2)
Wire nuts (yellow and red)
Tape (regular and colors)

Beyond this list of materials for interior wiring, there are a number of items that are needed for building services. They typically include the following:

Circuit-breaker type loadcenter (usually main circuit-breaker type)
Meter bases (100–200 amp)
Meter hubs (1½" and 2") and closures
1¼" cable connectors (indoor)
1¼" cable connectors (outdoor)
1½" PVC conduit, for 150-amp service riser (or as required by municipality or utility)
1½" PVC service head
1½" PVC terminal adaptor (connector)
1½" straps (2-hole)
2" PVC, for 200-amp service riser (or as required by municipality or utility)
2" PVC service head
2" PVC terminal adaptor
2" straps
2½" PVC (for underground services, line side)
2½" 90-degree elbows
2½" straps (2 hole)
#2/0 aluminum wire (for 150-amp services)
#4 aluminum wire (for 150-amp services)
#4/0 aluminum wire (for 200-amp services)
#2 aluminum wire (for 200-amp services)
#2/0 aluminum SE or SEU cable (for 150-amp services)
#4/0 aluminum SE or SEU cable (for 200-amp services)
#6 bare copper wire (150-amp grounding electrode conductor)
#4 bare copper wire (200-amp grounding electrode conductor)
Waterpipe clamp
Ground rods (⅝" x 8' galvanized)
Ground-rod clamps
Lightning arrestor (residential type)

3″ terminal adaptor (for branch circuit-cable entrances to the panel, where permitted)
3″ locknuts
½″ PVC conduit (misc. uses)
½″ PVC terminal adaptors
½″ locknuts
½″ straps (1-hole or 2-hole)
Circuit breakers:
1 pole,15 amp
1-P, 20 amp
2-P, 15 amp
2-P, 20 amp
2-P, 30 amp
2-P, 50 amp

JOB LAYOUT

In running an efficient residential wiring business, prejob planning is critical. The job plans should be expressly detailed to show each home run, and preferably the exact routing of each circuit. For repetitive houses (houses of which many will be built from the same set of working drawings), a detailed copy of the plan should be drawn on a sturdy grade of cardboard, so that the plans need only be detailed the first time. After the job is done, the plans should be filed away (somewhere you can find them again without too much trouble). This can save a lot of time if you ever have to go back to do any maintenance or service work. Remember that the better the instructions you give your electricians, the better their productivity is going to be.

STANDARDIZATION

Establish certain standards within your company as to specific installation techniques and dimensions. The standards that are most important for residential wiring are as follows:

1. *Exactly how three-way switches will be wired.* There are a number of different ways to wire three-way switches, and there is no standard method. If nothing else, you should specify a standard method and one or two alternate methods. When an alternate method is desired the plans should be specifically noted as to this fact.
2. *The marking of wires.* If your people won't be able to make all of their splices before the drywall is hung, you will need to have some method

of marking the wires so that your electricians will have some way of telling which wires are which. The wires that are especially important to mark are switchlegs, three-way travelers, and the ends of three-way switching circuits. If these wires are marked, hot wires and neutrals can be left alone, and the electrician will know that unless the wires are marked they are nothing more than hots and neutrals. Methods of marking typically include crimping the ends of the wires, twisting the wires, or coiling the wires or cables. The wires should also be tagged in the panel as to which circuit is which. This way there will be no need to identify circuits by testing with a continuity tester after the house is complete.

3. *The heights of receptacles and switches.* This is especially critical above countertops in kitchens. If these receptacles are not all mounted at the same height it will stand out like a sore thumb.

INSTALLATION PROCEDURE

By specifying an exact order for your electricians to follow in their installations, you guarantee that the work will be done in the most efficient manner. The following is the recommended procedure:

1. Unload all materials and tools that will be used that day. Place all materials and tools in an accessible and central location.
2. Run an extension cord to the power source.
3. Fill and hang the required cable dishes.
4. Mount all of the electrical boxes. Receptacles at 14″ OC (on center); switches at 48″ OC; outlets above counter tops at 42″ OC; use pancake boxes screwed to the bottom of ceiling joists for ceiling fans and light fixtures with only one cable running to them. Use round boxes for other light fixtures.
5. Mount plaster rings for low-voltage outlets (TV and telephone).
6. Drill all holes and notch any framing members as required. Use a 1″ auger bit with the right-angle drill for drilling studs and plates. Drill several large holes above loadcenter for home runs, more holes if required.
7. Install cable *exactly as shown on plans*; as each area is completed, highlight plans. Using stilts, install branch circuit cable first, then TV and phone wire, followed by power circuits (dryer, range, A/C units, heater, etc.). Leave 8–10 inches of cable coming out of all boxes. Run any special circuits last (post lights, well pumps, etc.). Mark all cables as required during the installation.
8. Install service panel. Pull knock-outs and install connectors before

mounting panel. Use a large terminal adaptor and locknut (3″) in top knock-out for bringing branch circuits into the panel (where approved). Run service cable.

9. Bring all cables into panel; be sure that all are marked. Use a connector for service cable. Cover the panel front with cardboard. This keeps paint and drywall mud off of your buss bars during the processes of finishing the walls.
10. Mount and wire meter base.
11. Drive ground rod and run grounding electrode conductor. Run grounding conductor from panel to waterpipe if required.
12. Install service riser for overhead service, or drop to underground service as required. Install lightning arrestor where required.
13. Splice all branch circuit cables.
14. Double check the entire job to be sure that everything is complete. Load all materials and tools back into the truck. Thoroughly clean up the job site.

SELLING AND COLLECTING

As a residential contractor, your customers will almost exclusively be builders of single-family residences and developers for large complexes. The first thing that a homebuilder will look for is a low or reasonable price, but after that, he will be concerned about service. Homebuilders are very time-conscious regarding the construction of their homes. This is because they take out a loan for the construction of the home and sometimes for the purchase of the land also. The longer the construction and sales process takes, the more interest the builder will have to pay. Builders would like all of their subcontractors to do their work quickly, professionally, and at a day's notice. In residential construction, the homebuilder and the electrical contractor should work as partners. Do your best to keep this kind of relationship with your local homebuilders. If they know that they can call you and get the job done properly with no delays your price doesn't have to be the absolute lowest.

Provide options to your homebuilders. Some of these options typically include intercom systems, security systems, low voltage switching, energy-management systems, remote-control systems, fire alarm systems, and the like. You can also offer different types of wiring devices and plates as an option. Sell *up* when you can.

Once a job or portion of a job is complete and billable, see that you bill it right away. If you bill your jobs immediately, residential work can give you an excellent cash flow. Usually it works best to bill 60 percent of the contract amount after the completion of the rough-in, and the remaining 40 percent along with any extra work as soon as the job is complete. If the builder submits his bills to the bank promptly he should get his draws in a number of days. This usually allows for the electrical contractor to get paid in about a week from the time the work is completed, which is excellent for the construction industry.

Large multifamily projects are almost always more competitive than single-family homes but allow your people to work in one place for a long time, which cuts down on the time spent setting up each individual project.

11

BUSINESS FACTORS

Even if a contractor could know absolutely everything that there is to know about electrical contracting (and no one really does), there are a number of other factors that will also play a large part in determining the company's success. These factors usually involve the running of a business, and extend way beyond the mechanics of electrical contracting and electrical installations. It is important to take care of these other factors in your company's success, as they don't just take care of themselves as most of us would like.

GETTING HELP FROM PROFESSIONALS

An electrical contractor isn't supposed to be an authority on construction law or accounting procedures or insurance. He is supposed to be a good electrical contractor. Certainly he should know enough about these subjects to make intelligent decisions, but an electrical contractor just doesn't have the time to be an expert on every pertinent business-related topic. It is a practical impossibility. When your company needs to make decisions on matters other than those that pertain to the electrical side of contracting, get some professional input, and when you do need this type of input, try to get the best people that you can. As with almost anything else, in the long run it will pay you to have the best.

Capital

One of the absolutely most important prerequisites for a contracting company is to have enough capital. This is one of the most regrettable facts about this business, but it *is* a fact, and it cannot be ignored. You cannot let your company take on more work than it can finance. The temptations to go beyond this point are strong, but remember that there are hundreds of contracting companies (probably thousands) that are no longer in business for no other reason than the fact that they did not have enough capital to keep their company going.

The first solution to this problem is to go to a bank to negotiate a line of credit. If you can find a bank willing to do this it is the best alternative, but banks can be very conservative and are often unwilling to make loans to small construction firms, especially newer firms.

The next solution (which very few people like) is to get a larger contractor to go into a joint venture with your company. This is virtually the same thing as taking on a partner, which has drawbacks, but half of something big is better than all of almost nothing.

The third solution to this problem is to do no more work than you can afford, no matter how small an amount that may be. Then save every bit of the profits that you can, and put them into a separate account until you have enough capital to take on a larger project. Then repeat this process until your company is adequately financed. The problem with this solution is that it requires far more self-control than most people have, and the process can take a number of years to build up any substantial amount of capital.

If you try the first alternative (which is also the most pleasant), don't give up after being rejected by the first bank you visit. Prepare a portfolio of your work and an itemized business plan. Make an impressive presentation to the bankers you call on, and call on as many as you have to. Different banks have different policies, and different banks have more or less money that they need to loan out. Keep on trying until you either succeed or run out of banks to call on.

Accounting

The processes of accounting and bookkeeping have become increasingly complex down through the years, and each time a new tax law is enacted or the IRS makes a rule change, the factors affecting bookkeeping and accounting methods change again.

For this reason it is essential that your company maintains a consistent relationship with a good accounting firm. Find a firm that handles a number of contracting companies, as they will be familiar with the factors that matter to your company. The best way to do this is to call on a number of other contractors in your area; it doesn't matter what type of contractor (general, mechanical, roofing), just that they are in the construction business. Ask these people frankly who their accountant is and if they are happy with the service they get. You will find most of these people quite willing to help out with information.

Beyond just the accounting of costs and expenses, there are a number of factors that determine the best way to set up your company's bookkeeping and accounting systems. There are a number of different ways, all with different advantages and disadvantages. Taxes, inventory, completed contracts, and contracts-in-progress all come into play, so it is crucial that you have input from an expert. Unless you have an accounting degree, there is almost no way that you can know all of the right answers. (To further complicate this, laws vary from state to state and area to area; and even these laws are subject to change every few years.)

CORPORATIONS AND LIABILITY

In the construction business, the possibility of large lawsuits and large losses are ever-present. The entire industry is by nature tumultuous. Things don't always go as they are planned, and some people don't do what they said they would do. This is especially dangerous because the entire industry is so interdependent on the individual parts. For example, let's assume that a large developer is constructing a large hotel; during the middle of the construction process, the progress payments start to lag; the developer explains that the bank was slow to process the payments because of recent holidays or some other nice-sounding excuse. After a couple more weeks of delays, someone runs into the office and informs you that the developer has just filed for bankruptcy, the job has been shut down, and the job-site is fenced-in and locked. Now you call the developer who refers you to his lawyer. The lawyer explains the situation, informs you that the matter is entirely in the hands of the judge, and says that there will be no money distributed (even if there is any left) for at least six months. So now what happens? The subcontractors have about 60 days-worth of progress payments coming to them which they cannot hope to get for at least six months, and more than likely they will get almost nothing even then. But the workers still expect their weekly paychecks, and the suppliers are still entitled to be paid for materials that they delivered to the subcontractors.

The end result of situations like this (and they are not uncommon) is that a number of subcontractors will go out of business and perhaps one or two suppliers with them. There will be a flurry of liens and judgments against both the developer and the subcontractors.

It is important to realize that in the construction business, someone can get put in a very bad legal position even if it is not his own fault, and even if he makes only a marginal error at the wrong time, the results can be disastrous. For this reason, it is prudent for the subcontractor to either incorporate or have extensive insurance. Because the costs of incorporating are fairly low and non-recurring, almost all subcontractors choose this solution. The corporation insulates the owner of the company from personal liability in the case of judgments against the company. Without this protection, the owner of the company could stand to lose his personal assets because of a problem over which he has no control.

There are other benefits to incorporating besides this, but nothing that can compare in importance. Of course there are decisions to make as to type of corporation, capitalization, stock, officers, and so on, and for these you will need the advice of both a good lawyer and an accountant. Consult with them before deciding on details, but by all means form a corporation if you have not already.

Insurance

Like forming a corporation, insurance has found a permanent place in the construction industry because of the excessive liabilities inherent to the industry.

Insurance comes in a multitude of types and styles. Each contracting company has its own specific needs, and as said earlier, it is essential to discuss your needs with a professional. Take care also that you deal with an insurance agent who is familiar with the construction industry and the industry's specific needs. An agent who is unfamiliar with the construction business may not even realize that you need certain types of coverage.

The three main types of insurance that electrical contractors typically need are general liability coverage, comprehensive automobile liability coverage, and Workmen's Compensation coverage. Since Workmen's Compensation coverage is required by law in every state, and regulations and requirements vary from state to state, we will exclude Workmen's Compensation from these discussions.

To start, it should be understood that insurance policies are contracts under which the insurer agrees to pay the insured (or a beneficiary designated by the insured) an agreed-upon sum of money, should certain events occur.

A comprehensive general liability policy will cover liability hazards known at the start of coverage and sometimes unknown hazards that may develop while the policy is in force. The problem with general liability policies is that they often do not cover (and often specifically exclude) many liabilities that contractors are subject to. This makes it necessary for many contractors to get supplemental coverage. The types of supplemental coverage often required for contracting companies are contractual or hold-harmless coverage, completed-operations coverage, broad-form property damage coverage, and explosion, collapse, and underground coverage (XCU).

Hold-harmless clauses are frequently found in construction contracts; many contractors and subcontractors sign these contracts and thereby bind themselves to the clauses without really knowing what they mean. These hold-harmless clauses require the contractor to protect a project owner, architect, and general contractor from any losses caused by their company's work. This means that if one of your company's workers makes a mistake, your company is responsible not only for any damages, but also for everyone else's legal fees and judgments. So, one wrong move by one of your company's people could wipe out your company. *Broad-form* hold-harmless policies are available which can cover this liability.

Completed-operations coverage is important because a general liability policy covers losses while the contractor is on the job, but not after. A completed-operations policy fills this gap.

Broad-form property damage coverage is often required because of exclusions in general liability policies which exclude the insured from collecting on damage to another person's property while working on it or with it. These exclusions also stop the insured from collecting on damage to the insured's own property. In other words, if your company is working on a customer's generator and your men damage it, a regular liability policy will not cover the damages, but the broad-form property damage policy will.

XCU coverage (explosion, collapse, and underground) is of importance mainly to contractors who do a lot of underground work. It can be more expensive than other types of coverage, but for contractors doing a large amount of this work it can be very important.

MANAGEMENT SUCCESSION

One of the more interesting facts about the construction business is that only about one in five of the construction companies doing business today will survive into the next generation. This means that once the owner of the company either dies or retires, the company will go out of business in a short period of time.

The real core of this problem is that most construction companies are run almost entirely by one person who has kept the company going by carefully handling every detail personally. It is very difficult for this type of individual to let go a little bit and let someone else (who maybe has a little different style) take over some of the work. Once the individual is gone, the company has little chance of long-term survival, as no one else really knows all of the ins and outs of the company's operations.

The next generation of management must be trained to do the whole job if the company is to survive. Unfortunately, this is not a very pleasant task for the older generation who is presently in control. Generations differ because the forces and events that mold one generation's outlook a certain way are vastly different from the forces that mold the next generation's outlook. This makes for different outlooks on the company's operations, and sometimes makes for a different set of values altogether. Letting someone with ideas different from your own run your company is not an easy thing to do. It requires the boss to resign from being boss and to take the role of company coach.

Other factors that come into play here are the values of the company, lifestyles, estate planning, getting equity out of the company, and others. For these questions you will need the help of a professional who can deal with the complexities of your individual company.

Investments

Outside investments are an important part of an electrical contractor's personal financial planning. The fact is that most electrical contractors who are wealthy didn't make the bulk of their money in electrical contracting; they made it in outside investments. Their electrical-contracting businesses gave these contractors a steady income and provided an extra chunk of cash every now and then, which is typical of almost every electrical contracting business (and for every other type of contracting also). By its nature, the contracting market is cyclical. During slow times a contractor does well to keep his company going and to pay all of his bills, but when times are good, there will be extra money to do something with.

The question that faces the electrical contractor is what to do with this money when it comes in. One answer is to reinvest the money in the company, usually by purchasing new tools or equipment. While this idea definitely has merit, it isn't always in the owner's best interests. The next answer is for the owner to invest some of the profits in some type of outside investment. This has a number of benefits to the contractor.

1. It diversifies the contractor's assets and risks; that is, the contractor doesn't put all of his eggs in one basket. If the contractor doesn't do this and something happens to the company, all will be lost. The construction business is a far more risky business than many others available for a contractor to put money in, and is a fairly dangerous place to put all of your assets.
2. It provides a way for the contractor to "hide" a little bit of money for himself. Most contractors don't take very much money out of their companies. By using a little of the excess profits for personal investments every now and then, the contractor is securing his future.
3. Many outside investments are more profitable than electrical contracting. The average electrical-contracting company makes only about a three-to-five percent annual profit. Almost any outside investment can do better than this.
4. Many types of outside investments are very well suited for an electrical contractor. Investments require planning and care as do contracting companies. By virtue of the electrical contractor being in a managerial position, he can take a few minutes here and there to check up on his investments without any trouble.

Real estate. Real-estate investments have historically been the best for electrical contractors. There are several reasons why real estate is especially good for electrical contractors, but it must always be remembered that the state of the real-estate market is the crucial factor. Regardless of how convenient real-estate investments are for the electrical contractor, if the real-estate market offers poor returns the contractor should look elsewhere for investment opportunities. The benefits that real estate offers to electrical contractors are usually these:

1. The little chunks of profits that come in every now and then are perfect for use as a down-payment on a rental property. If the rental property is good, the monthly rent will cover the loan payments and maintenance (or at least come very close). By doing this, the contractor invests several thousands of dollars to control a property worth many times that much, while the tenants (over the course of time) buy the building for him.
2. Maintenance of rental properties is very cheap and easy for an electrical contractor. If something needs to be fixed, the odds are about ten to one that a good electrician can fix it. By sending his own employee to fix the problem, the contractor gets the problem fixed right away for just the cost of wages.
3. The contractor has a central office with bookkeeping and clerical capabilities. This makes the job of managing rental properties (and other types of investments also) very convenient and inexpensive. He can also use his other employees to do a lot of his running around; for example, by having the superintendent stop by the property and make sure that the new tenants have moved in.

Real estate offers the additional benefit that it has appreciated more than any other type of investment over the last several decades. Also, it can give more leverage than almost any other type of investment with a good safety factor. (Leverage is using a small amount of money to control something worth a large amount of money, like using a down-payment of $10,000 to control an apartment building worth $140,000).

Always work with professionals in any type of investment, real estate or otherwise. Never get into a transaction without the advice of a lawyer who specializes in real-estate law.

Stocks, bonds, and commodities. Stocks, bonds, and commodities can offer terrific returns, high leverage, and sometimes a low initial investment. But even with these advantages, they are not for someone who is at all unfamiliar with these markets. Almost any electrical contractor can walk into an apartment building and know whether it is well built, but very few electrical contractors are experts (or anything close to an expert) at the commodities markets. The electrical contractor understands buildings and people better than precious-metal markets, and is usually better off sticking with something he understands.

Nonetheless, there is a case to be made for these other investments *but* the contractor will need the continual advice of a competent professional who is trustworthy and loyal beyond question.

The subcontractor spec-house idea. This is the best investment idea for smaller contractors (and larger ones too) that the author has ever heard, provided that you can assemble a group of subcontractors who are willing to work together (they are more willing to work together than they usually are when they do it to make money, as is the case here). The process is this:

1. Assemble a group of subcontractors, a general contractor, and usually a developer (enough trades to complete a single-family home).
2. All agree to and sign documents (drawn up by a lawyer) stating that each contractor will provide his service for his specialty of work for the house. Each contractor will at this time itemize the exact value of his work. The general contractor will provide his services, and the developer (or other) will provide the building lot. All values will be agreed upon and put in writing at this time.
3. Each trade performs their work until the house is complete. Then a realtor is contracted and the house is sold.
4. The proceeds of the sale are distributed to each partner in the venture according to the value of their products and services. For example, the electrical contractor's services for wiring the house come to $2400 and the total value of all partners in the venture's services come out to $40,000. If the house then sells for $60,000, each partner gets a return of the value of his services, plus 50 percent. The electrical contractor will get a check for $3600; that's not bad for a small investment over a couple of months.

The great benefit of this idea is two-fold for even a small contractor. It gives the contractor a way to make an outside investment without it hurting, and as long as the local real-estate market is fairly good, the amount of risk is minimal.

12

HOW THE GAME IS REALLY PLAYED

Anytime and in any profession that money comes into the picture, things go on that shouldn't occur. People bend the rules, people play favorites, sometimes they even break the law, all in the pursuit of money. Even if you know exactly what the right thing to do is, and the right time to do it, there are other factors that come into play, factors that don't follow the rules.

This chapter is an explanation of inner dealings within the electrical-contracting industry. These are all actual events within the author's personal experience.

BIDDING TACTICS

There is probably no end to the number of tricks contractors use to try to figure out who is bidding what and how to protect their own prices while finding out their competitors' prices. Here are some of the ways they try to do it:

1. *The undercut technique.* By simply working with a general contractor who is willing to tell the electrical contractor the other electrical prices, the general contractor hopes to get the lowest price. He thinks that the electrical contractor he favors will undercut any other electrical price he gives him. It is notable that a lot of electrical contractors who have tried this trick have gotten burned. The reason: Any general contractor who is unethical enough to go along with this will not think twice about quoting a false price to the electrical contractor. The GC tells the electrical contractor a price that is several thousand dollars below the true low bid; the electrical contractor then beats that price by a couple of thousand dollars and gets the job. But when he tries to do the job, he ends up short. As one old-timer aptly put it: "If a man will lie, he'll steal too."
2. *The bait and switch technique.* This one is used primarily by reputable contractors who are trying to protect themselves from unethical competition. Here's how it works: In almost any area there are general contractors who shop prices, even before bid time. The reputable electrical contractor calls in a bid to one of these unethical general contractors. He makes the bid high, and calls it in early, knowing that the general contractor will tell this price to other electrical contractors, and that these other contractors will undercut his first price by a couple of thousand dollars. The good electrical contractor will then call in the right price (about ten thousand dollars below the first

price) to reputable general contractors, and at the last minute to the unethical GC.

3. *The out-of-town fake.* This is similar to bait and switch; but for jobs where there are a number of contractors (both GC and electrical) from another town bidding on the project. The electrical contractor calls the out-of-town general contractors with a high price. These general contractors will more often than not relate these prices to their local electrical subs, who beat this price by a little, and call it in to the general contractors in both cities. The first electrical contractor then gives his real price to only his local general contractors, making him the low bidder.
4. *The outright deception.* This is the worst one of them all, and yes, it does happen. The culprit in this case is a very unethical electrical contractor. He pretends that he is a general contractor, and calls up several electrical contractors to try to solicit bids from them. The ones that do it can sound very authentic, even giving a phone number and address (phony of course) to the electrical contractor, and discussing intricate parts of the job to allude to their authenticity.

Bid Teams

This is one tactic that is both fair and honest, yet gives the contractor some advantage over the competition. In this instance a group of subcontractors team together and bid their combined prices to general contractors. In other words, an electrical contractor, mechanical contractor, roofer, and drywaller combine all of their individual prices into one lump sum, which they will call in to the general contractors. If all of the individual prices are competitive, the team will have a pretty good percentage of winning bids. To the best of the author's knowledge, this is perfectly legal and ethical.

Supplier Tricks

Believe it or not, electrical suppliers often have just as hard a time making money as electrical contractors do. And electrical contractors and their suppliers play the same kind of tricks between themselves as electrical contractors and general contractors. Here are several things to look out for:

1. *Billings.* Suppliers often have to cut their percentage to the bone to get orders. Sometimes they try to make up for this by splitting up their billing of a large order into several different invoices, which are often sent at different times. The total of these invoices will add up to more than the amount the electrical contractor and supplier agreed

upon originally. Because of the way the billing is split up, the discrepancy often slips by the electrical contractor.

2. *Codes and numerals.* Suppliers need to use some type of numbering system to keep their charges and inventory in order, but sometimes they take advantage of this by using numbers or codes with no description on invoices. They then charge a higher than normal price which they hope will go unnoticed.

The Art of Accounting

Don't always believe what an accountant says. There is an art to accounting; a good accountant can change a company's books around any number of ways and yet they can be perfectly legal. This can't go on forever, but for a year or two a company's books can be made to look a lot better or worse than they really are.

Some companies even keep two sets of books, one made to look like the company is making almost no money for the IRS's eyes, and one set that shows the company as being very profitable to show to bankers and bonding companies. Apparently there is some way of doing this, and still staying within the bounds of the law (however loosely interpreted it may be).

Avoiding Blame

To some of the upper echelons in the construction industry the name of the game seems to be "don't ever get caught making a mistake." Architects, engineers, and manufacturers are particularly bad about this. You can save yourself trouble if you remember this fact, and try to work around it.

Be careful of purchase orders and purchase-order confirmations from manufacturers and suppliers. Sometimes they can have such tricky wording that it almost takes a lawyer to make sense out of them. Rest assured that this wording is not going to be in your favor.

Manufacturer's Reps

Some of the most notorious people in the business are the manufacturer's reps. (In all fairness, there are many, many manufacturer's reps that are honest and ethical, but the facts are that some of the most devious tricks in the business are pulled off by manufacturer's reps; especially lighting reps.)

These are some of the tricks that have been pulled:

1. *Riding with one contractor.* What happens in this instance is that the rep will be working up a price for a large job, usually for light fix-

tures, and puts out a "street price;" that is, a high price that he quotes to all of his local distributors. Then, just before bid time, he will call the one or two contractors who have agreed to show him some loyalty, and give them a much better price, leaving a little bit of markup in the price for the distributor, which the rep will choose later. This gives the one or two contractors with the low price a tremendous advantage over the other contractors.

2. *Securing exclusive packages.* Again, this applies particularly to lighting reps, although not exclusively. By servicing and selling to specifiers (engineers and architects) reps are sometime able to get specifications written so that their lines of products are the only ones that will meet the specifications. This is especially effective for light-fixture packages. If the rep can get even one fixture specified that only he can meet, he will be able to quote on only a fixture *package* (meaning that the light fixtures will only be sold as a complete package, and not individually), and because of the one fixture he will be the only person who can properly supply the order. The last detail is that the price goes way up.
3. *Changing catalog numbers and products.* Sometimes reps and manufacturers will play games with catalog numbers; sometimes only with prefixes and suffixes. They can then substitute different grades of materials from the ones that were supposed to have been specified, but only as long as the construction process won't be too highly scrutinized.

There is a way to beat a rep at his own game sometimes. If a rep is trying to lock up a package of light fixtures or some other type of material (as explained in the second point), you can sometimes have a friendly contractor in a different territory buy the order for you (at a more reasonable price), and have it drop-shipped directly to your job site. However, this doesn't always work, as sometimes reps in different territories talk to each other frequently and protect each other's sales.

Front-Loading

Front-loading is a technique that some electrical contractors use to try to improve their cash flow. They charge more for the first several billings on a project than they are really worth. They then have extra cash to work with, which helps them finance their jobs. There is one big problem with this: Sooner or later things will have to even up. By getting extra money out of the beginning of the job, you are going to get less than you should out of the end of the job, which can really hurt.

Dealing with Personalities

The greatest advantages that are available to contractors come in knowing how to handle and understand different people. Even with the age of computers and legally worded documents, personalities play a decisive role and will continue to do so for the forseeable future.

It is people who make the decisions that control the contracting industry. People with different strengths and flaws, different likes and dislikes, and different pressures and requirements facing them.

A contractor who knows the people in his industry can often anticipate their actions. This gives the contractor just enough of an edge to get a job that ordinarily would have gone to someone else. Or perhaps a contractor can anticipate a problem coming, and get out of the way before it hits.

This is a skill that takes time to develop, but if you watch old-time contractors long enough, you will notice that this is how they get an advantage over younger contractors.

INDEX

C

D

H

I

J

L

O

P

T

U

W

X

Y